HYDROBORATION

HYDROBORATION

HERBERT C. BROWN

R. B. Wetherill Research Professor
Purdue University

W. A. BENJAMIN, INC.   new york   1962

To my wife —
her patience, forbearance, and understanding
made the writing of this book
a pleasure

Preface

Only a few short years have passed since we first observed that olefins could be simply and quantitatively converted into organoboranes. This development opened up an almost virgin area for exploration. The research has been fruitful and progress has been rapid. As a consequence, it has become increasingly difficult for the research man not actively involved in research in this area to maintain contact with the developments.

This fact has been brought home to me in observing the difficulties experienced by my new postdoctorates and graduate students who wished to participate in this research program. In spite of careful guidance and direction, they found it necessary to devote considerable time and energy to tracing down individual papers and specialized reviews widely scattered throughout the literature.

If this was the situation with men who were devoting full time to this research area, what must be the difficulties of research workers who might wish to follow the developments in this area, but could afford to devote only a fraction of their available time to this activity! Yet the hydroboration–organoborane area is not a specialized field of investigation that the average research chemist can safely ignore. It is developing into a major synthetic tool, as important as the Grignard reagent, condensations, and the complex hydrides to the chemist who utilizes synthetic procedures in his research.

It appeared important, therefore, to make the developments available in easily assimilable form for graduate students and research chemists. With considerable reluctance I decided to tear myself away from the exciting current research and to undertake the preparation of a systematic review. My task was facilitated by the various reviews I had previously published in this area, and I made considerable use of these surveys as well as of the individual papers we have published exploring the characteristics of the hydroboration reaction.

The hydroboration area and the chemistry of organoboranes is still in a stage of rapid development. Consequently, it did not appear desirable to prepare a detailed reference work. Instead, I have aimed at presenting a complete summary of the results to date, with emphasis on its readability. To this end, I adopted an informal approach, presenting the observed phenomena and my interpretations in an informal manner, much as I should do in classroom lectures on the subject. I have also utilized chemical equations and structures liberally, in an effort to provide easy, visual comprehension of the versatile reactions and changes made possible by this new approach.

Unexpectedly, it proved to be a real pleasure to write this book. I hope that the reader will experience equal pleasure in becoming acquainted with a new, fascinating area of chemistry.

HERBERT C. BROWN

Lafayette, Indiana
September 28, 1961

Contents

HYDROBORATION

1 | Introduction and Survey

The observation that olefins may be readily converted into organoboranes under mild experimental conditions, first reported in 1956 and 1957, has provided a major new route to these interesting derivatives.[1,2]

$$9RCH{=}CH_2 + 3NaBH_4 + AlCl_3 \longrightarrow 3(RCH_2CH_2)_3B + AlH_3 + 3NaCl$$

$$12RCH{=}CH_2 + 3NaBH_4 + 4BF_3 \longrightarrow 4(RCH_2CH_2)_3B + 3NaBF_4$$

$$6RCH{=}CH_2 + B_2H_6 \longrightarrow 2(RCH_2CH_2)_3B$$

Not only does the boron-hydrogen bond add rapidly and quantitatively to carbon-carbon double and triple bonds, but it adds also with remarkable ease to carbon-oxygen double bonds[3] and to carbon-nitrogen double and triple bonds.[4]

[1] H. C. Brown and B. C. Subba Rao, *J. Am. Chem. Soc.*, **78**, 5694 (1956).

[2] H. C. Brown and B. C. Subba Rao, *J. Org. Chem.*, **22**, 1136 (1957).

[3] H. C. Brown, H. I. Schlesinger, and A. B. Burg, *J. Am. Chem. Soc.*, **61**, 673 (1939).

[4] H. C. Brown and B. C. Subba Rao, *J. Org. Chem.*, **22**, 1135 (1957).

Consequently, the addition of the boron-hydrogen linkage to multiple bonds between carbon and carbon, nitrogen, or oxygen appears to be a reaction of very wide generality, as general as the addition of hydrogen, but proceeding under much milder conditions. The term hydroboration, with the symbol HB, has been proposed for this reaction. The present discussion will emphasize the hydroboration of olefinic and acetylenic derivatives and the reactions of the resulting organoboranes of interest in synthetic chemistry.

Early History

Early observations on the reaction of diborane with olefins indicated that the reaction was quite slow, requiring elevated temperatures and long reaction periods.[5] Reexamination of the reaction of simple olefins with diborane has revealed that the reaction is indeed slow, but is catalyzed remarkably by ethers.[6] Indeed, in ether solvents the reaction of diborane with representative olefins proceeds at a rate that is far too fast to follow by the usual kinetic techniques. The new hydroboration procedures are carried out in ether solvents, commonly diglyme (dimethylether of diethyleneglycol) (DG), tetrahydrofuran (THF), or ethyl ether (EE). Presumably it was the inadvertent use of these solvents that was primarily responsible for the first quantitative reactions realized in the newer procedures (Chapter 2).

Chemistry of Organoboranes

For many years the only practical route to the organoboranes involved the reaction of an organometallic, generally the organozinc[7] or the organomagnesium[8] derivative, with the boron ester or halide. Since the synthesis of the organoborane involved the prior formation of an organometallic, there was little incentive in exploring the possible utility of the organoboranes, another class of organometallics, as intermediates, in organic synthesis. Consequently, until recently, there was little information available on reactions of interest to the synthetic chemist.[9]

For example, it was known that the aliphatic organoboranes are typical nonpolar substances, which in their physical properties resemble closely the

[5] D. T. Hurd, *J. Am. Chem. Soc.*, **70**, 2053 (1948).

[6] H. C. Brown and L. Case, unpublished observations.

[7] E. Frankland and B. F. Duppa, *Proc. Roy. Soc. (London)*, **10**, 568 (1859).

[8] E. Krause and R. Nitsche, *Ber.*, **54**, 2784 (1921).

[9] M. F. Lappert, *Chem. Revs.*, **56**, 959 (1956).

corresponding hydrocarbons. They are relatively stable to water, but are sensitive to oxygen. They react sluggishly with the halogens and with the halogen acids, but the reactions do not appear to be of synthetic value.

$$R_3B + O_2 \xrightarrow{\text{H}_2\text{O}} R_2BOR$$

$$R_3B + O_2 \xrightarrow{\text{dry}} RB(OR)_2$$

$$R_3B + Br_2 \longrightarrow R_2BBr + RBr$$

$$R_3B + HBr \longrightarrow R_2BBr + RH$$

The organoboranes are coordinatively unsatisfied. They readily form addition compounds with ammonia and the amines.

$$R_3B + NR'_3 \rightleftharpoons R_3B \colon NR'_3$$

Although this reaction has proved of major value in the investigation of steric effects,[10] it also does not appear to be of significance to the synthetic chemist.

Indeed, only the ready oxidation of organoboranes with alkaline hydrogen peroxide, briefly mentioned by Johnson and Van Campen,[11] appeared to offer utility to the synthetic chemist.

$$R_3B + 3H_2O_2 + NaOH \longrightarrow 3ROH + NaB(OH)_4$$

Now that the organoboranes are readily synthesized from olefins without an organometallic as intermediate, there has developed new interest in exploring the chemistry of the organoboranes. Many new reactions of major significance to synthetic chemistry have been discovered (Chapter 3).

Thus the oxidation of organoboranes by alkaline hydrogen peroxide has been developed into a convenient procedure for converting olefins via the organoboranes into the corresponding alcohols.[12]

[10] H. C. Brown, *J. Chem. Soc.*, **1956**, 1248.
[11] J. R. Johnson and M. G. Van Campen, Jr., *J. Am. Chem. Soc.*, **60**, 121 (1938).
[12] H. C. Brown and B. C. Subba Rao, *J. Am. Chem. Soc.*, **81**, 6423 (1959).

The reaction is essentially quantitative and appears to be general. It proceeds cleanly, without any evidence of rearrangement, placing a hydroxyl group at the precise position occupied by the boron atom in the organoborane. This reaction has been of immense value in studying both the directive effects in hydroboration (Chapter 7) and the stereochemistry of the hydroboration reaction (Chapter 8).

The organoboranes are relatively stable to aqueous acids and bases, but they undergo protonolysis by carboxylic acids with surprising ease. Consequently, hydroboration followed by protonolysis offers a convenient route for the hydrogenation of double bonds.[13]

$$RCH{=}CH_2 \xrightarrow{\ HB\ } RCH_2CH_2{-}B\diagup \xrightarrow{\ CH_3CO_2H\ } RCH_2CH_3$$

Relatively little attention has yet been devoted to the utilization of organoboranes as Grignard-like reagents, for the transfer of alkyl groups to electron-deficient centers. However, Honeycutt and Riddle have successfully synthesized diethylmercury and tetraethyllead by treating the metal oxides with triethylborane and sodium hydroxide in water.[14]

$$2(C_2H_5)_3B + 3HgO + 3H_2O + 2NaOH \longrightarrow 3(C_2H_5)_2Hg + 2NaB(OH)_4$$

It is evident that such a transfer of alkyl groups from boron to other elements in aqueous systems should be an exceedingly valuable process in synthetic chemistry.

Treatment of the trialkylborane with silver nitrate in the presence of sodium hydroxide brings about a rapid coupling reaction. In this way, an olefin such as 2-methyl-1-pentene is readily converted into 4,7-dimethyldecane in yields of 75 to 80 per cent.[15]

[13] H. C. Brown and K. Murray, *J. Am. Chem. Soc.*, **81**, 4108 (1959).
[14] J. B. Honeycutt, Jr. and J. M. Riddle, *J. Am. Chem. Soc.*, **81**, 2593 (1959).
[15] H. C. Brown and C. H. Snyder, *J. Am. Chem. Soc.*, **83**, 1001 (1961).

This ready formation of carbon-carbon bonds under such mild conditions should find important application in organic synthesis. By use of two different olefins it is possible to build up many different carbon structures. Moreover, since many different functional groups can be accommodated in the hydroboration reaction, this opens up the possibility of a general synthesis of carbon structures, both with and without functional groups.[16]

Finally, there is evidence that this reaction proceeds via free radicals.[17]

$$2R-B{\textstyle<} + 2CCl_4 \xrightarrow[\text{NaOH}]{\text{AgNO}_3} 2RCl + C_2Cl_6$$

Consequently, hydroboration provides a promising new entry into the chemistry of free radicals.

Although relatively little has yet been published on the hydroboration of functional derivatives (Chapter 19), Hawthorne and Dupont realized a very promising result in their study of the hydroboration of allyl chloride.[18] The product, a γ-chloroorganoborane, undergoes rapid cyclization, on treatment with aqueous alkali, to form cyclopropane.

This is a highly promising new route to cyclopropane derivatives.

Isomerization of Organoboranes

At moderate temperatures the organoboranes undergo a facile isomerization which proceeds to place the boron atom at the least hindered position of the alkyl groups[19,20] (Chapter 9).

[16] H. C. Brown, C. Verbrugge, and C. H. Snyder, *J. Am. Chem. Soc.*, **83**, 1001 (1961).

[17] H. C. Brown, N. C. Hébert, and C. H. Snyder, *J. Am. Chem. Soc.*, **83**, 1002 (1961).

[18] M. F. Hawthorne and J. A. Dupont, *J. Am. Chem. Soc.*, **80**, 5830 (1958).

[19] H. C. Brown and B. C. Subba Rao, *J. Org. Chem.*, **22**, 1136 (1957).

[20] G. F. Hennion, P. A. McCusker, E. C. Ashby, and A. J. Rutkowski, *J. Am. Chem. Soc.*, **79**, 5190 (1957).

$$
\begin{array}{ccccc}
\begin{array}{c}
CH_3 \\
| \\
CH_2 \\
| \\
CH \\
\parallel \\
CH \\
| \\
CH_2 \\
| \\
CH_3
\end{array}
&
\xrightarrow{\ HB\ }
&
\begin{array}{c}
CH_3 \\
| \\
CH_2 \\
| \\
CH\!-\!\!B\!\!< \\
| \\
CH_2 \\
| \\
CH_2 \\
| \\
CH_3
\end{array}
&
\xrightarrow[1\ hr]{160^\circ}
&
\begin{array}{c}
CH_2\!-\!\!B\!\!< \\
| \\
CH_2 \\
| \\
CH_2 \\
| \\
CH_2 \\
| \\
CH_2 \\
| \\
CH_3
\end{array}
\end{array}
$$

This makes possible a number of very valuable syntheses, not otherwise practical. For example, the following syntheses may be readily achieved in essentially quantitative yield by utilizing hydroboration and isomerization.[21]

Displacement Reactions of Organoboranes

The mechanism of the isomerization reaction appears to involve a partial dissociation of the organoborane into olefin and the boron-hydrogen bond, followed by readdition. The process occurs repeatedly, until the boron atoms end up at the least hindered position of the molecule, thereby yielding the most stable of the organoboranes derivable from the particular alkyl groups used (Chapter 10).

[21] H. C. Brown and G. Zweifel, *J. Am. Chem. Soc.*, **82**, 1504 (1960).

It follows from this mechanism that the presence of another olefin, of equal or greater reactivity, will cause the original olefin to be displaced from the organoborane.[19,22,23] The combination of hydroboration, isomerization, and displacement makes possible the contrathermodynamic isomerization of olefins.

[22] R. Köster, *Ann.*, **618**, 31 (1958).
[23] H. C. Brown and M. V. Bhatt, *J. Am. Chem. Soc.*, **82**, 2074 (1960).

Borohydride Chemistry

In large part the developments in the hydroboration area were made possible by major improvements in the synthesis of diborane and alkali metal borohydrides.[24,25] Thus the reaction of lithium hydride with boron trifluoride in ethyl ether provided a convenient synthesis of diborane. Moreover, the reaction of methyl borate with sodium hydride at 250° made sodium borohydride readily available. Finally, the reaction of sodium borohydride with boron trifluoride etherate in diglyme solution provided a simple, quantitative synthesis of diborane.

$$6LiH + 2BF_3 \xrightarrow{\text{EE}} B_2H_6 + 6LiF$$

$$4NaH + B(OCH_3)_3 \xrightarrow{250°} NaBH_4 + 3NaOCH_3$$

$$3NaBH_4 + 4BF_3 \xrightarrow{\text{DG}} 2B_2H_6 + 3NaBF_4$$

These developments have taken these chemicals out of their previous category as laboratory rarities and have made them available as laboratory and industrial reagents of considerable versatility (Chapter 4).

Hydroboration Procedures

With this background, it proved possible to develop a large number of different procedures for the hydroboration of olefins (Chapter 5). The most convenient procedures still are based on sodium borohydride or diborane itself.[2,26] However, it proved possible to circumvent the use of any particular solvent, or any specific hydride reagent. In place of sodium borohydride or lithium borohydride, one can utilize the simple hydrides.[26]

$$3RCH{=}CH_2 + 3LiH + BF_3 \longrightarrow (RCH_2CH_2)_3B + 3LiF$$

$$12RCH{=}CH_2 + 12NaH + 16BF_3 \longrightarrow 4(RCH_2CH_2)_3B + 12NaBF_4$$

[24] H. I. Schlesinger and H. C. Brown, in collaboration with B. Abraham, A. C. Bond, N. Davidson, A. E. Finholt, J. R. Gilbreath, H. Hoekstra, L. Horvitz, E. K. Hyde, J. J. Katz, J. Knight, R. A. Lad, D. L. Mayfield, L. Rapp, D. M. Ritter, A. M. Schwartz, I. Sheft, L. D. Tuck, and A. O. Walker, *J. Am. Chem. Soc.*, **75**, 186 (1953).

[25] H. C. Brown and P. A. Tierney, *J. Am. Chem. Soc.*, **80**, 1552 (1958).

[26] H. C. Brown, K. J. Murray, L. J. Murray, J. A. Snover, and G. Zweifel, *J. Am. Chem. Soc.*, **82**, 4233 (1960).

Lithium aluminum hydride has been utilized.[27] Finally, at elevated temperatures the amine-boranes may be utilized.[28,29]

$$12RCH{=}CH_2 + 3LiAlH_4 + 4BF_3 \longrightarrow 4(RCH_2CH_2)_3B + 3LiAlF_4$$

$$3RCH{=}CH_2 + R_3'N{:}BH_3 \xrightarrow{\;100{-}200°\;} (RCH_2CH_2)_3B + R_3'N$$

Scope of the Hydroboration Reaction

The hydroboration reaction has been demonstrated to be widely applicable (Chapter 6). Alkenes containing two, three, or four alkyl substituents on the double bond readily undergo hydroboration. Cyclic and bicyclic olefins, such as 1,2-dimethylcyclopentene and α-pinene, readily react. Aryl groups on the double bond are accommodated, as in 1-phenylcyclohexene, 1,1-diphenylethylene, *trans*-stilbene, and triphenylethylene.[30] Finally, a number of steroids with highly hindered double bonds have been demonstrated to react.[31] Consequently, the hydroboration reaction appears to be fully as applicable as the addition of bromine to the carbon-carbon double bond.

The great majority of olefins undergo complete reaction to form the corresponding trialkylborane. However, in the case of olefins with a large

[27] S. Wolfe, M. Nussim, Y. Mazur, and F. Sondheimer, *J. Org. Chem.*, **24,** 1034 (1959).

[28] M. F. Hawthorne, *J. Org. Chem.*, **23,** 1788 (1958).

[29] E. C. Ashby, *J. Am. Chem. Soc.*, **81,** 4791 (1959).

[30] H. C. Brown and B. C. Subba Rao, *J. Am. Chem. Soc.*, **81,** 6428 (1959).

[31] W. J. Wechter, *Chem. & Ind. (London)*, **1959,** 294.

degree of steric hindrance, the reaction appears to proceed rapidly only to the dialkylborane or monoalkylborane stage.[30]

This partial alkylation of diborane makes available a series of mono-alkylboranes and dialkylboranes, of interest as selective hydroborating agents (Chapters 13 and 14).

Directive Effects

The hydroboration of simple terminal olefins, such as 1-hexene, proceeds to place the boron predominantly on the terminal position. Consequently, the hydroboration of such olefins, followed by the oxidation *in situ* of the resulting organoborane, provides a very convenient procedure for the anti-Markownikoff hydration of double bonds.[12]

$$RCH{=\!=}CH_2 \xrightarrow{\text{HB}} RCH_2CH_2B{<} \xrightarrow{\text{[O]}} RCH_2CH_2OH$$

A more detailed study of directive effects in the hydroboration of representative olefins[32] (Chapter 7) reveals that the simple 1-alkenes, such as 1-pentene and 1-hexene, react to place 94 per cent of the boron on the terminal position, 6 per cent becoming attached at the 2 position. This distribution is not influenced significantly by branching of the alkyl group.

$$CH_3CH_2CH_2CH{=\!=}CH_2$$
$$\uparrow \quad \uparrow$$
$$6\% \quad 94\%$$

$$H_3C-\underset{\underset{H_3C}{|}}{\overset{\overset{H_3C}{|}}{C}}-CH{=\!=}CH_2$$
$$\uparrow \quad \uparrow$$
$$6\% \quad 94\%$$

An aryl substituent, as in styrene, causes increased substitution in the nonterminal position.

$$\langle\rangle\!-\!CH{=}CH_2$$
$$\uparrow \quad \uparrow$$
$$20\% \quad 80\%$$

It is interesting that the distribution can be altered considerably by substituents in the *para* position.

An alkyl substituent in the 2 position favors attachment of the boron to the terminal carbon atom.

[32] H. C. Brown and G. Zweifel, *J. Am. Chem. Soc.*, **82**, 4708 (1960).

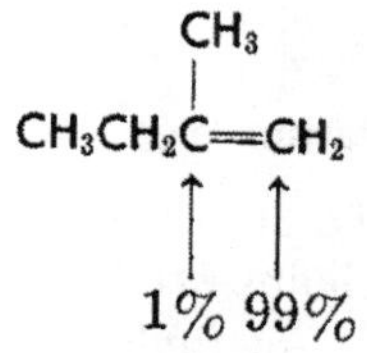

$$CH_3$$
$$|$$
$$CH_3CH_2C{=}CH_2$$

$\uparrow$ $\uparrow$

1% 99%

A similar preference for the less substituted position is demonstrated in internal olefins.

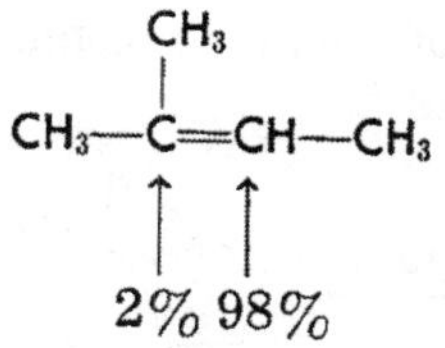

$$CH_3$$
$$|$$
$$CH_3{-}C{=}CH{-}CH_3$$

$\uparrow$ $\uparrow$

2% 98%

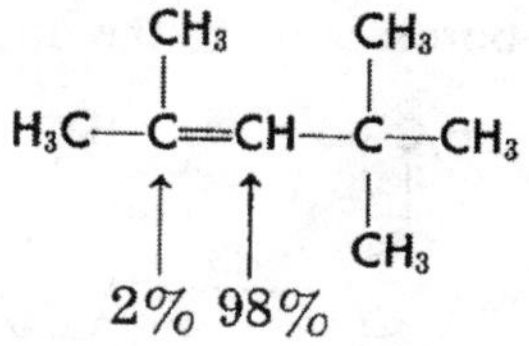

$$CH_3 \qquad CH_3$$
$$| \qquad\quad |$$
$$H_3C{-}C{=}CH{-}C{-}CH_3$$
$$\qquad\qquad\qquad |$$
$$\qquad\qquad\qquad CH_3$$

$\uparrow$ $\uparrow$

2% 98%

There is no significant discrimination between the two positions of an internal olefin containing groups of markedly different steric requirements.

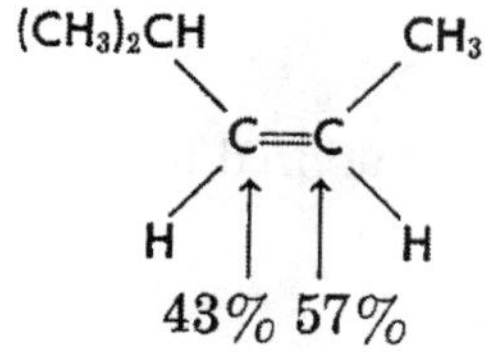

$$(CH_3)_2CH \qquad CH_3$$
$$\diagdown \qquad\qquad \diagup$$
$$C{=}C$$
$$\diagup \qquad\qquad \diagdown$$
$$H \qquad\qquad\qquad H$$

$\uparrow$ $\uparrow$

43% 57%

(However, it has been possible to achieve selective hydroboration of this olefin at the less hindered position by the use of the reagent, bis-3-methyl-2-butylborane, disiamylborane, Chapter 13.)

It is concluded that the hydroboration reaction involves a simple four-center transition state, with the direction of addition controlled primarily by the polarization of the boron-hydrogen bond $\left(\underset{}{\overset{\delta^+ \ \ \delta^-}{{>}B{-}H}}\right)$.

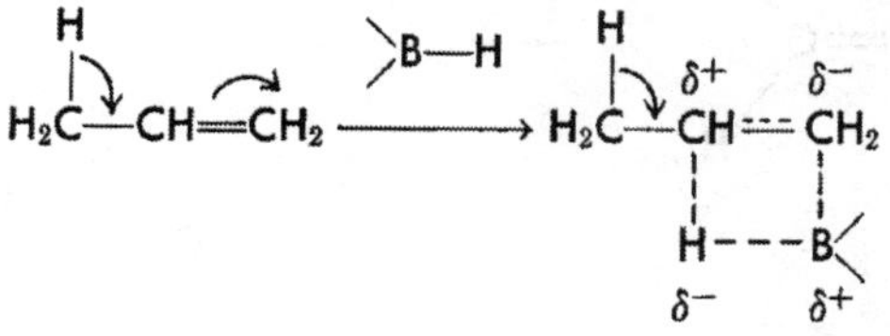

$$\underset{}{\overset{H}{|}}$$
$$H_2C{-}CH{=}CH_2 \quad \overset{{>}B{-}H}{\longrightarrow} \quad H_2C\overset{H}{\underset{}{|}}CH\overset{\delta^+}{=\!=\!=}CH_2$$
$$\qquad\qquad\qquad\qquad\qquad\qquad\qquad H{-}{-}{-}B\!\!<$$
$$\qquad\qquad\qquad\qquad\qquad\qquad\quad \delta^- \qquad \delta^+$$

Stereochemistry of Hydroboration

This four-center transition state is likewise supported by the results of the hydroboration of cyclic olefins[33] (Chapter 8). Thus the hydroboration of 1-methylcyclopentene and 1-methylcyclohexene, followed by oxidation with alkaline hydrogen peroxide, results in the formation of pure *trans*-2-methylcyclopentanol and *trans*-2-methylcyclohexanol. Since the hydrogen peroxide oxidation evidently proceeds with retention of configuration (Chapter 3), the hydroboration reaction must involve a *cis* addition of the hydrogen-boron bond to the double bond of the cyclic olefin.

This *cis* hydration has been utilized to achieve a convenient synthesis of diastereomeric alcohols.[34]

[33] H. C. Brown and G. Zweifel, *J. Am. Chem. Soc.*, **81**, 247 (1959).
[34] E. L. Allred, J. Sonnenberg, and S. Winstein, *J. Org. Chem.*, **25**, 26 (1960).

The hydroboration of norbornene proceeds to give *exo*-2-norborneol almost exclusively.

The generalization has been proposed that hydroboration proceeds by way of a *cis* addition, preferentially from the less hindered side of the double bond.[33] This generalization is now supported by a large number of independent observations. For example, α-pinene is converted into isopinocampheol[33] and β-pinene into *cis*-myrtanol.[35]

It is interesting that heat isomerizes the borane from β-pinene to the less hindered tri-*trans*-myrtanylborane. Oxidation of the isomerized organoborane produces *trans*-myrtanol.[36] This result clearly demonstrates that the hydroboration stage is controlled by the reaction mechanism and not by the stability of the product.

Finally, the hydroboration of cholesterol produces cholestane-3β, 6α-diol preferentially, clearly the result of a preferential *cis* addition from the less hindered underside of the molecule.[37]

[35] H. C. Brown and G. Zweifel, *J. Am. Chem. Soc.*, **83**, 2544 (1961).
[36] J. C. Braun and G. S. Fisher, *Tetrahedron Letters*, No. **21**, 9 (1960).
[37] W. J. Wechter, *Chem. & Ind. (London)*, **1959**, 294.

The clean stereochemical results achieved with the hydroboration reaction promise to make this synthetic route of major value for the chemist interested in the synthesis of stereochemically defined products.

Hydroboration of Hindered Olefins

It was pointed out earlier that the reaction of hindered olefins with diborane appears to proceed rapidly only to the dialkylborane stage in some cases, and to the monoalkylborane stage in others. A detailed study of the hydroboration of hindered olefins has demonstrated that the reaction can be utilized for the synthesis of a number of mono- and dialkylboranes in high purity and essentially quantitative yield[38] (Chapter 11).

Thus the reaction of 2-methyl-2-butene at 0° readily produces bis-3-methyl-2-butylborane (disiamylborane), and the reaction of the more hindered olefin, 2,4,4-trimethyl-2-pentene, is readily controlled to form 2,4,4-trimethyl-3-pentylborane.

[38] H. C. Brown and A. W. Moerikofer, *J. Am. Chem. Soc.*, in press.

Similarly, the reaction of the tetrasubstituted olefin, 2,3-dimethyl-2-butene, can be directed quantitatively to the formation of the corresponding monoalkylborane, 2,3-dimethyl-2-butylborane (thexylborane).

In the case of the simple cyclic olefins, the reaction of cyclopentene proceeds rapidly to the tricyclopentylborane stage. However, the lower reactivity of cyclohexene, as well as the insolubility of the intermediate, dicyclohexylborane, permits the ready synthesis of this derivative.

Both 1-methylcyclohexene and α-pinene are easily converted into the corresponding dialkylboranes.

Finally, the addition of 2-methyl-1-propene, 2-methyl-1-butene, 2,4,4-trimethyl-1-pentene, and β-pinene to an equivalent quantity of borane in tetrahydrofuran at 0° gives 60 to 70 per cent of the dialkylboranes, 10 per cent of free borane, and 20 per cent of the corresponding trialkylborane.

60–70%

Alkylboranes

The ready availability of these mono- and dialkylboranes via the hydroboration reaction prompted a detailed study of their physical and chemical properties[39] (Chapter 12). Both molecular-weight determinations (in ethyl ether and tetrahydrofuran) and infrared examination reveal that both disiamylborane and monothexylborane exist as the dimers. Consequently, they should be considered derivatives of diborane.

Both sym-tetrasiamyldiborane and sym-dithexyldiborane add diborane reversibly at 0° to form 1,1-disiamyldiborane and monothexyldiborane, respectively. The reaction proceeds much further toward completion in the case of sym-tetrasiamyldiborane, presumably because the large steric strains in this derivative are largely relieved in the formation of the product.

$$(Sia_2BH)_2 + (BH_3)_2 \rightleftharpoons 2Sia_2BH_2BH_2$$

$$(t\text{-}HexBH_2)_2 + (BH_3)_2 \rightleftharpoons 2t\text{-}HexHBH_2BH_2$$

sym-Tetrasiamyldiborane reacts with trimethylamine to form an addition compound which is stable at low temperatures but is largely disso-

[39] H. C. Brown and G. J. Klender, *J. Inorg. Chem.*, in press.

ciated into its components at room temperature. *sym*-Dithexyldiborane forms an addition compound which is stable at room temperature.

$$(Sia_2BH)_2 + 2(CH_3)_3N \rightleftharpoons 2Sia_2BH:N(CH_3)_3$$

$$(t\text{-}HexBH_2)_2 + 2(CH_3)_3N \longrightarrow 2t\text{-}HexBH_2:N(CH_3)_3$$

The treatment of borane in tetrahydrofuran at 0° with one equivalent of cyclohexene ($C_6H_{10}/BH_3 = 1$) results in the predominant formation of 1,1-dicyclohexyldiborane. This observation suggests the following reaction path for hydroborations in tetrahydrofuran.

In this solvent diborane exists as the monomeric species, tetrahydrofuran-borane. The first step in the hydroboration reaction must involve the addition of this component to the olefin to give the corresponding monoalkylborane.

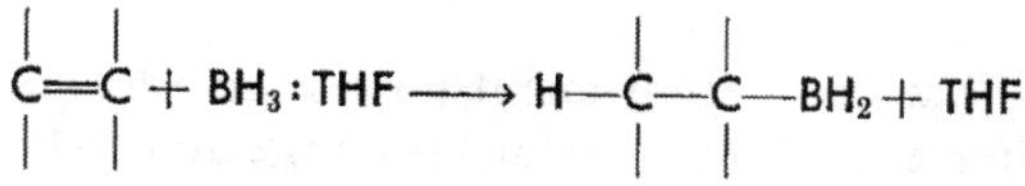

$$\overset{|}{\underset{|}{C}}=\overset{|}{\underset{|}{C}} + BH_3:THF \longrightarrow H-\overset{|}{\underset{|}{C}}-\overset{|}{\underset{|}{C}}-BH_2 + THF$$

In the case of highly hindered olefins, such as 2,3-dimethyl-2-butene and 2,4,4-trimethyl-2-pentene, a second reaction to place an additional alkyl group becomes relatively slow, and the monoalkylborane moieties dimerize to form the *sym*-dialkyldiboranes, which may be identified in the reaction solution and isolated therefrom.

However, with less hindered olefins, such as cyclohexene, the reaction with the monoalkylborane is faster than the reaction with tetrahydrofuran-borane. Consequently, in such cases the monoalkylborane is converted into the dialkylborane, in spite of the presence of free tetrahydrofuran-borane. The results show that the dialkylborane is capable of removing the borane group from tetrahydrofuran.

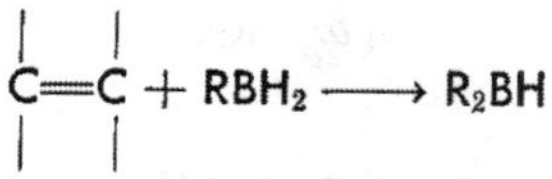

$$\overset{|}{\underset{|}{C}}=\overset{|}{\underset{|}{C}} + RBH_2 \longrightarrow R_2BH$$

$$R_2BH + H_3B:THF \rightleftharpoons R_2BH_2BH_2 + THF$$

Continued reaction with olefin produces the tetraalkyldiborane and, finally, the trialkylborane, in the case of the less hindered derivatives.

Selective Hydroborations with Disiamylborane

The observation that the hydroboration of 2-methyl-2-butene proceeds rapidly to the dialkylborane stage, but only very slowly to the trialkylborane end product, indicates that the last stage must be strongly influenced by the large steric requirements of both the reagent and the olefin.

$$
2 \quad \underset{\substack{\text{H}_3\text{C} \quad \text{H}}}{\overset{\substack{\text{H}_3\text{C} \quad \text{CH}_3}}{\text{C}=\text{C}}} + \text{BH}_3 \xrightarrow{\text{fast}} \text{H}-\underset{\substack{\text{H}_3\text{C} \quad \text{H}}}{\overset{\substack{\text{H}_3\text{C} \quad \text{CH}_3}}{\text{C}-\text{C}}}-)_2\text{BH}
$$

$$
\text{H}-\underset{\substack{\text{H}_3\text{C} \quad \text{H}}}{\overset{\substack{\text{H}_3\text{C} \quad \text{CH}_3}}{\text{C}-\text{C}}}-)_2\text{BH} + \underset{\substack{\text{H}_3\text{C} \quad \text{H}}}{\overset{\substack{\text{H}_3\text{C} \quad \text{CH}_3}}{\text{C}=\text{C}}} \xrightarrow{\text{slow}} \text{H}-\underset{\substack{\text{H}_3\text{C} \quad \text{H}}}{\overset{\substack{\text{H}_3\text{C} \quad \text{CH}_3}}{\text{C}-\text{C}}}-)_3\text{B}
$$

This suggested that the reagent might possess an enhanced sensitivity to the steric requirements of the substituents on the double bond and thereby exert a steric influence on the direction of hydroboration and on the relative reactivities of various olefin structures[40] (Chapter 13).

Indeed, 1-hexene reacts readily with the reagent, and oxidation of the product yields 1-hexanol in an isomeric purity of at least 99 per cent, whereas diborane yields 6 per cent of the secondary alcohol. Similarly, styrene is converted by this procedure into β-phenylethyl alcohol of 98 per cent purity in contrast to the 80:20 distribution realized with diborane. Finally, cis-4-methyl-2-pentene undergoes attachment of the boron atom to the less hindered position, providing the alcohol 4-methyl-2-pentanol in isomeric purity of 97 per cent.

$$\text{CH}_3\text{CH}_2\text{CH}_2\text{CH}=\text{CH}_2$$

1% 99%

2%   98%

[40] H. C. Brown and G. Zweifel, *J. Am. Chem. Soc.*, **82**, 3222 (1960).

$$3\% \quad 97\%$$

A major development along this line is the recent report that treatment of Δ^1-cholestene with disiamylborane results in the predominant formation of cholestan-2α-ol, in contrast to the nearly 50:50 mixture of cholestan-1α-ol and cholestan-2α-ol realized with diborane.[41]

These studies revealed major differences in the rates of reaction of the reagent with different olefins. Thus the reaction of the reagent with 1-hexene and 2-methyl-1-pentene is very rapid at 0°, whereas the reactions of internal olefins are considerably slower, with cyclopentene reacting faster than cis-2-hexene and the latter reacting considerably faster than cyclohexene. Cis-2-hexene reacts considerably faster than the trans isomer. Trisubstituted olefins, such as 2-methyl-2-butene and 1-methyl-cyclohexene, react very slowly.

The result of these qualitative experiments may be expressed in the following series[42]: 1-hexene ⩾ 3-methyl-1-butene > 2-methyl-1-butene > 3,3-dimethyl-1-butene > cyclopentene ⩾ cis-2-hexene > trans-2-hexene > trans-4-methyl-2-pentene > cyclohexene ⩾ 1-methylcyclopentene > 2-methyl-2-butene > 1-methylcyclohexene ⩾ 2,3-dimethyl-2-butene.

The observed differences in reactivity are quite large and can be utilized for the selective hydroboration of a more reactive olefin in the presence of a less reactive structure. Thus, treatment of a mixture of 1-pentene and 2-pentene with sufficient reagent to react with the more reactive terminal

[41] F. Sondheimer and M. Nussim, J. Org. Chem., 26, 630 (1961).
[42] H. C. Brown and G. Zweifel, J. Am. Chem. Soc., 83, 1241 (1961).

olefin yields the 2-pentene free of the 1- isomer.   In the same way, the 1-hexene, in a mixture with cyclohexene, can be selectively hydroborated to yield essentially pure cyclohexene.   It is also possible to react selectively cyclopentene in a mixture with cyclohexene.   Finally, it is possible to take advantage of the differences in reactivity to remove the more reactive *cis* olefin from the *trans* isomer.

It is apparent from these results that the reagent should permit the selective hydroboration of complex molecules containing two or more reactive centers.   Some applications of this kind will be discussed in Chapter 15.

Although the reaction of diborane with olefins is too fast for measurement, it is proved possible to make a kinetic study of the reaction of disiamylborane with a number of olefins.[43]   The reaction in tetrahydrofuran at 0° is second order, first order in disiamylborane dimer, and first order in olefin.

The rate constant for cyclopentene, $k_2 = 14.0 \times 10^{-4}$ liter mole^{-1} sec^{-1}, decreases by a factor of 100-fold for cyclohexene, $k_2 = 0.134 \times 10^{-4}$ liter mole^{-1} sec^{-1}, but rises sharply again by a factor of 500 for cycloheptene, $k_2 = 72 \times 10^{-4}$ liter mole^{-1} sec^{-1}.   The rate constant for *cis*-2-butene, with $k_2 = 23 \times 10^{-4}$, is greater than that of *trans*-2-butene, $k_2 = 3.8 \times 10^{-4}$, by a factor of 6.

The results suggest that strain in the olefin system greatly facilitates the reaction.   Thus both cyclopentene and cycloheptene are strained olefins relative to cyclohexene.   Similarly, *trans*-2-butene is thermodynamically more stable than the *cis* isomer.

Finally, the second-order kinetics appear to require a transition state composed of one molecule of the dimer and one molecule of olefin.   The reaction presumably proceeds with the formation of the product and one molecule of monomer, which should react with a second molecule of olefin in a rapid second step, or undergo dimerization to *sym*-tetrasiamyldiborane.

$$Sia_2BC_5H_{11} + Sia_2BH$$

[43] H. C. Brown and A. W. Moerikofer, *J. Am. Chem. Soc.*, **83**, 3417 (1961).

The ether solvent probably facilitates the reaction by coordinating with the disiamylborane monomer, the "leaving group" in this reaction.

Asymmetric Hydroborations with Diisopinocampheylborane

As was pointed out earlier, the hydroboration of α-pinene proceeds readily to the formation of diisopinocampheylborane.

Since α-pinene is available from natural sources in optically active form, this reaction makes available an optically active dialkylborane. This reagent exhibits a remarkable ability to achieve the asymmetric hydroboration of suitable olefins[44] (Chapter 14).

For example, cis-2-butene reacts with the reagent from α-pinene, $[\alpha]_D^{20} + 47.6°$, to produce an organoborane which, oxidized with alkaline hydrogen peroxide, produces 2-butanol, $[\alpha]_D^{20} - 11.8°$. This rotation corresponds to an optical purity of 87 per cent. Use of laevorotatory α-pinene produces 2-butanol of the opposite rotation, $[\alpha]_D^{20} + 11.8°$.

[44] H. C. Brown and G. Zweifel, J. Am. Chem. Soc., 83, 486 (1961).

Similarly, *cis*-3-hexene was converted in 81 per cent yield to 3-hexanol, $[\alpha]_D^{20} - 6.5°$, indicating an optical purity of 91 per cent. Application of the procedure to norbornene produces *exo*-norborneol in a yield of 62 per cent. The observed rotation, $[\alpha]_D^{20} - 2.0°$, also indicates a high optical purity.

These results establish that a boron atom at the asymmetric center, RR'C*HB, is capable of maintaining asymmetry without significant racemization over periods of several hours. The unusually high optical purities realized in this asymmetric hydroboration and the ease with which organoboranes may be converted into other derivatives with retention of configuration promise to make this approach to optically active derivatives a useful one for the synthetic chemist.

In the three cases mentioned, the alcohols from dextrorotatory α-pinene have configurations which can be correlated with the structure of the reagent. Consequently, the procedure also promises to be valuable as a tool to establish the absolute configuration of alcohols and other derivatives.

Hydroboration of Dienes

The hydroboration of dienes, followed by oxidation with alkaline hydrogen peroxide, provides a convenient route to the corresponding glycols. In some cases the monohydroboration of dienes can be controlled to produce the unsaturated alcohol[45] (Chapter 15).

Thus, 1,5-hexadiene is converted into 1,6-hexanediol, and 2,3-dimethylbutadiene is converted into 2,3-dimethyl-1,4-butanediol.[46]

$$
\begin{array}{l}
H_2CCH{=}CH_2 \quad HB \quad [O] \quad H_2CCH_2CH_2OH \\
| \qquad\qquad \longrightarrow \longrightarrow \qquad | \\
H_2CCH{=}CH_2 \qquad\qquad\qquad H_2CCH_2CH_2OH
\end{array}
$$

$$
\begin{array}{l}
\quad CH_3 \quad CH_3 \quad HB \quad [O] \quad H_3C \quad CH_3 \\
\quad\; | \qquad | \qquad \longrightarrow \longrightarrow \qquad\; | \qquad | \\
H_2C{=}C{-\!-\!-}C{=}CH_2 \qquad\qquad HOH_2CCHCHCH_2OH
\end{array}
$$

Application of this procedure to 1,3-butadiene results in the formation of a mixture of diols — approximately 30 per cent 1,3- and 70 per cent 1,4-butanediol. It would appear that the same factors which operate in styrene to place 20 per cent of the boron in the secondary position cause

[45] H. C. Brown and G. Zweifel, *J. Am. Chem. Soc.*, **81**, 5832 (1959).
[46] G. Zweifel, K. Nagase, and H. C. Brown, *J. Am. Chem. Soc.*, **84**, 183 (1962).

30 per cent of the boron to become attached to the nonterminal position of the butadiene system.

$$H_2C{=}CH{-}CH{=}CH_2 \xrightarrow{\ HB\ } H_2C{-}CH_2{-}CH_2{-}CH_2 \quad 70\%$$

with B groups on the first and last carbons

$$+ \ H_3C{-}CH{-}CH_2{-}CH_2 \quad 30\%$$

The situation is considerably improved by the use of disiamylborane for the dihydroboration of olefins of this kind.[47]

The dihydroboration of cyclic dienes has received relatively little attention as yet. However, Saegebarth has reported that the dihydroboration of cyclopentadiene with excess diborane provided a 41 per cent yield of cis-1,3-cyclopentanediol.[48]

Dienes, such as 1,3-butadiene, react with diborane to produce polymeric organoboranes. For example, treatment of borane in tetrahydrofuran at 0° with 1,3-butadiene in the ratio to form the cyclic derivative

or

[47] G. Zweifel, K. Nagase, and H. C. Brown, *J. Am. Chem. Soc.*, **84,** 190 (1962).
[48] K. A. Saegebarth, *J. Org. Chem.*, **25,** 2212 (1960).

results instead in the formation of a polymeric product with a molecular weight in the neighborhood of 350. The product reacts readily with methanol to form the corresponding ester, but the latter is not volatile. It therefore appears that the dominant reaction in the hydroboration of dienes with diborane at 0° is not cyclization to form simple cyclic derivatives,[49,50] but involves instead the formation of polymeric derivatives which may contain cyclic units within the structure. The simple cyclic structures reported by Köster appear to arise as a result of thermal rearrangement of the initial reaction products.

The isomerization of the organoboranes derived from dienes offers interesting synthetic possibilities.[51] The hydroboration-oxidation of 1,3-pentadiene yields a mixture — 10 per cent 1,3- and 90 per cent 1,4-pentanediol. However, heating of the reaction product under isomerization conditions results in the production of essentially pure 1,5-pentanediol. On the other hand, isomerization of the organoborane from 1,5-hexadiene produces 70 per cent 1,5- and 30 per cent 1,6-hexanediol. It therefore appears that the isomerization proceeds to the preferred formation of a six-membered boron heterocycle.

The monohydroboration of conjugated aliphatic dienes is a difficult problem. The conjugated dienes are less reactive than the corresponding olefins. Consequently, the monohydroboration of such a diene converts it into an unsaturated borane which is far more reactive toward the hydroborating agent than the original diene. As a result, the yield of monohydroborated product is low.

The situation is less difficult in the case of nonconjugated dienes. Thus 1,5-hexadiene has been converted into 5-hexene-1-ol and bicycloheptadiene has been converted into exo-5-norbornenol in reasonable yield.

[49] R. Köster, *Angew. Chem.*, **71**, 520 (1959).
[50] R. Köster, *Angew. Chem.*, **72**, 626 (1960).
[51] K. A. Saegebarth, *J. Am. Chem. Soc.*, **82**, 2081 (1960).

Disiamylborane would appear to have major advantages in the mono-hydroboration of dienes with two reaction centers of considerably different reactivities. Thus, both 4-vinylcyclohexene and d-limonene are readily converted into unsaturated alcohols by this reagent, with attack at the more reactive exocyclic double bond.

Even in the case of conjugated dienes, it has proved possible to achieve a good yield of the unsaturated alcohol by taking advantage of the sensitivity of disiamylborane toward different olefinic structures.

Finally, interesting and useful results have been realized in the mono-hydroboration of terpenes, such as myrcene and caryophyllene, with this reagent.[52]

Hydroboration of Acetylenes

The hydroboration of acetylenes proceeds readily, with results of considerable utility for the chemist interested in synthesis[53] (Chapter 16).

[52] H. C. Brown and K. P. Singh, unpublished observations.
[53] H. C. Brown and G. Zweifel, *J. Am. Chem. Soc.*, **81**, 1512 (1959).

The treatment of internal acetylenes, such as 3-hexyne, with the theoretical quantity of hydroborating agent results in the formation of the corresponding trivinylborane. The resulting products are readily converted into the corresponding *cis* olefins in high purity by treatment with acetic acid at 0°.

$$
\begin{array}{c}
C_2H_5 \\
| \\
C \\
\parallel \\
C \\
| \\
C_2H_5
\end{array}
+ \ BH_3 \ \longrightarrow \
\begin{array}{c}
C_2H_5-C-H \\
\parallel \\
C_2H_5-C-)_3B
\end{array}
\xrightarrow{\ HOAc\ } 3
\begin{array}{c}
C_2H_5-C-H \\
\parallel \\
C_2H_5-C-H
\end{array}
$$

$$
\begin{array}{c}
C_6H_5 \\
| \\
C \\
\parallel \\
C \\
| \\
C_6H_5
\end{array}
+ \ Sia_2BH \ \longrightarrow \
\begin{array}{c}
C_6H_5-C-H \\
\parallel \\
C_6H_5-C-BSia_2
\end{array}
\xrightarrow{\ HOAc\ }
\begin{array}{c}
C_6H_5-C-H \\
\parallel \\
C_6H_5-C-H
\end{array}
$$

The corresponding reaction of 1-alkynes with diborane is less simple. The reaction appears to proceed preferentially to the dihydroboration stage. However, disiamylborane circumvents this difficulty. The product is readily converted to the 1-alkene with acetic acid.

$$
n\text{-}C_4H_9C\!\!\equiv\!\!CH + Sia_2BH \longrightarrow n\text{-}C_4H_9C\!\!=\!\!CH
$$
$$
\begin{array}{cc}
| & | \\
H & BSia_2
\end{array}
$$
$$
\downarrow HOAc
$$
$$
n\text{-}C_4H_9CH\!\!=\!\!CH_2
$$

Oxidation of the vinylboranes with alkaline hydrogen peroxide converts them into the corresponding carbonyl derivatives.

$$
C_2H_5C\!\!\equiv\!\!CC_2H_5 \xrightarrow{HB} \xrightarrow{[O]} C_2H_5CH_2CC_2H_5
$$
$$
\parallel
$$
$$
O
$$

$$
n\text{-}C_4H_9C\!\!\equiv\!\!CH \xrightarrow{HB} \xrightarrow{[O]} n\text{-}C_4H_9CH_2CHO
$$

It is evident that this ready conversion of terminal acetylenes to the corresponding aldehyde should find useful application in synthetic work.

The dihydroboration of acetylenes appears to place two boron atoms on the same carbon atom.[54] However, hydrolysis of one of the boron-carbon bonds appears to occur rapidly during the oxidation procedure. As a result, the product consists of a mixture of both alcohol and carbonyl derivative, unless special precautions are taken.

Diborane as a Reducing Agent

Diborane is a powerful reducing agent, which readily reduces many groups[3,4,55] (Chapter 17). However, it was indicated earlier that the speed of the reaction of diborane with many double and triple bonds is such that it should be possible to achieve the synthesis of organoboranes with a wide variety of functional groups. This opens up the possibility of syntheses via the organoboranes which are not feasible with other types of organometallics. Before discussing the hydroboration of unsaturated derivatives containing such functional groups, it appears desirable to review the behavior of diborane as a reducing agent. It should be understood that the products indicated are obtained after hydrolysis of the boron-containing intermediate.

Both aliphatic and aromatic aldehydes and ketones are rapidly reduced to alcohols at room temperature. However, the presence of electronegative substituents appears to reduce the reactivity of these groups toward the reagent — both chloral and acid chlorides are relatively stable toward reduction. γ-Lactones and oxides are reduced readily to the corresponding glycol and alcohol. Esters are reduced to alcohols, but the reaction is considerably slower than the reductions previously considered. The sodium salts of carboxylic acid do not appear to undergo reduction under these conditions. Similarly, both nitro compounds and sulfones appear to resist reduction. Perhaps the most unexpected development is the observation that both carboxylic acids and nitriles are rapidly reduced by the reagent at room temperature, yielding alcohols and amines, respectively. Alkyl and aryl halides, thioethers, and ethers do not react.

Tertiary amines form addition compounds but do not undergo reduction. Alcohols and phenols evolve hydrogen but are not otherwise attacked.

An examination of the relative reactivity of a number of groups toward diborane indicates the following order of decreasing reactivity.[56] The

[54] H. C. Brown and G. Zweifel, *J. Am. Chem. Soc.*, **83**, 3834 (1961).

[55] H. C. Brown and B. C. Subba Rao, *J. Am. Chem. Soc.*, **82**, 681 (1960).

[56] H. C. Brown and W. Korytnyk, *J. Am. Chem. Soc.*, **82**, 3866 (1960).

groups below the line do not exhibit any significant reactivity toward reduction by this reagent.

> carboxylic acids
> olefins
> ketones
> nitriles
> epoxides
> esters

> acid chlorides
> nitro compounds
> sulfones
> alkyl and aryl halides
> thioethers
> ethers
> alcohols and phenols

The marked difference in the relative reactivities of these groups toward sodium borohydride and diborane has both important theoretical and practical implications (Chapter 17). However, the subject will not be considered at this point because it is not germane to the question under consideration. Since so many of the groups exhibit either no significant activity toward diborane or are less reactive toward the reagent than olefins themselves, it is evident that it should be possible to achieve the hydroboration of olefins in the presence of such groups.

Disiamylborane as a Reducing Agent

The introduction of the highly branched siamyl groups into borane profoundly modifies its reactivity toward olefins and acetylenes. These groups also have a significant effect upon the reducing properties of the reagent[57] (Chapter 18). Thus the reagent reduces only a limited number of groups at 0°. The products indicated are those obtained after hydrolysis of the intermediate boron derivative. Aldehydes and ketones are reduced to alcohols. Both γ-lactones and N,N-dimethylamides take up but one equivalent of disiamylborane. The γ-lactones produce the corresponding hydroxyaldehyde and the N,N-dimethylamides provide the related aldehydes. Nitrobenzene and nitriles react only very slowly with the reagent under these conditions.

[57] H. C. Brown and D. B. Bigley, *J. Am. Chem. Soc.*, **83**, 486 (1961).

The reagent reacts with the following groups to evolve hydrogen, but no reduction occurs: alcohols, phenols, carboxylic acids, amides, and sulfonic acids. No reaction has been observed with esters, acid chlorides, acid anhydrides, azobenzene, sulfones, and sulfonyl chlorides.

The reactivity of olefinic structures toward the reagent varies over such a wide range that it is not possible to state generally that every possible olefin will undergo selective hydroboration in the presence of these groups. However, terminal olefins are highly reactive toward the reagent. There is little doubt that it should be possible to hydroborate such olefins selectively ($RCH{=}CH_2$, $R_2C{=}CH_2$) without difficulty. Even in the case of internal olefins, no difficulty should be experienced with the great majority of groups.

Diisopinocampheylborane also exhibits interesting possibilities as a selective reducing agent.[58]

Hydroboration of Functional Derivatives

It is appropriate at this time to consider the application of the reagent to the hydroboration of functional derivatives (Chapter 19).

The presence of relatively inert substituents, such as halogen and ether groupings, should not cause any difficulty in the hydroboration reaction. For example, the hydroboration of p-chlorostyrene and p-methoxystyrene proceeds normally.[32] Similarly, the aliphatic derivatives, allyl chloride[18] and vinylethyl ether, have been successfully hydroborated.[59]

Even for cases in which the molecule contains reducible groups, it has been possible to achieve hydroboration. Thus methyl oleate has been converted to an equimolar mixture of 9- and 10-hydroxyoctadecanoic acids by the hydroboration-oxidation procedure[60] and the methyl ester of 10-undecenoic acid has likewise been converted into the 11-hydroxy derivative.[61]

It was pointed out that the free carboxylic acid group exhibits extraordinary reactivity toward diborane. Consequently, the direct hydroboration of 10-undecenic acid would appear to be impractical. Fortunately, it is possible to circumvent this difficulty with disiamylborane. This reagent reacts with the carboxylic acid group, evolving hydrogen, but fails to reduce it under hydroboration conditions.[57] Consequently, by

[58] H. C. Brown and D. B. Bigley, *J. Am. Chem. Soc.*, **83**, 3166 (1961).

[59] B. M. Mikhailov and T. A. Shchegoleva, *Izvest. Akad. Nauk. SSSR*, 546 (1959).

[60] S. P. Fore and W. G. Bickford, *J. Org. Chem.*, **24**, 920 (1959).

[61] R. Dulou and Y. Chrétien-Bessière, *Bull. soc. chim. France*, **1959**, 1362.

means of disiamylborane, 10-undecenoic acid is readily converted to 11-hydroxyundecanoic acid.

$$CH_2\!\!=\!\!CH(CH_2)_8CO_2H + 2Sia_2BH \longrightarrow Sia_2BCH_2CH_2(CH_2)_8CO_2BSia_2$$

$$\downarrow [O]$$

$$HOCH_2CH_2(CH_2)_8CO_2H$$

From these early results, it appears that the hydroboration of functional derivatives should offer no unusual difficulties. It may be anticipated, however, that the presence of substituents will markedly influence the direction of the hydroboration reaction, and this is a subject that as yet has received relatively little attention.

For example, the hydroboration of styrene proceeds to place 20 per cent of the boron on the secondary position of the side chain. The presence of a *p*-methoxy group decreases this to 9 per cent, whereas a *p*-chloro substituent raises it to 35 per cent.

Similarly, the hydroboration of *t*-butylethylene proceeds to give 6 per cent of the secondary and 94 per cent of the primary derivative. However, the related compound, trimethylvinylsilane, undergoes hydroboration to yield 37 per cent of the secondary derivative.[62]

[62] D. Seyferth, *J. Inorg. & Nuclear Chem.*, 7, 152 (1958).

These effects can be very large. Thus, it has been reported that the hydroboration of 3-chlorocyclohexene proceeds exclusively to form the 2-chlorocyclohexylborane.[63] Since compounds containing halogen in the 2 position undergo a rapid elimination reaction, the hydroboration does not lead to useful derivatives.

It is obviously desirable to have a systematic study of the influence of substituents on the direction of hydroboration. Such a study is underway.[64]

Related Reactions

The present volume is concerned with the hydroboration reaction and its implications for organic synthesis. However, it may not be inappropriate to review briefly certain related reactions.

In 1947 it was observed that trichlorosilane could be added to 1-octene under the influence of acetyl peroxide to yield n-octyltrichlorosilane.[65,66]

$$n\text{-}C_6H_{13}CH{=}CH_2 + HSiCl_3 \xrightarrow[50\text{--}60°]{(RCOO)_2} n\text{-}C_6H_{13}CH_2CH_2SiCl_3$$

[63] P. Binger and R. Köster, *Tetrahedron Letters*, No. 4, 156 (1961).

[64] H. C. Brown and K. Keblys, research in progress.

[65] L. H. Sommer, E. W. Pietrusza, and F. C. Whitmore, *J. Am. Chem. Soc.*, **69**, 188 (1947).

[66] E. W. Pietrusza, L. H. Sommer, and F. C. Whitmore, *J. Am. Chem. Soc.*, **70**, 484 (1948).

Later that year it was reported that the addition could be achieved without the added peroxide by heating the two components under pressure to temperatures of 160 to 400°.[67] The reaction was shown to be applicable to monomethyldichlorosilane and dimethylmonochlorosilane, as well as to trichlorosilane.

$$\underset{|\quad|}{\overset{|\quad|}{C}}{=}\underset{|}{\overset{|}{C}} + HSiCl_3 \xrightarrow{\sim 300°} H{-}\underset{|}{\overset{|}{C}}{-}\underset{|}{\overset{|}{C}}{-}SiCl_3$$

$$\underset{|\quad|}{\overset{|\quad|}{C}}{=}\underset{|}{\overset{|}{C}} + HSi(CH_3)Cl_2 \xrightarrow{\sim 300°} H{-}\underset{|}{\overset{|}{C}}{-}\underset{|}{\overset{|}{C}}{-}Si(CH_3)Cl_2$$

$$\underset{|\quad|}{\overset{|\quad|}{C}}{=}\underset{|}{\overset{|}{C}} + HSi(CH_3)_2Cl \xrightarrow{\sim 300°} H{-}\underset{|}{\overset{|}{C}}{-}\underset{|}{\overset{|}{C}}{-}Si(CH_3)_2Cl$$

The addition of the silicon-hydrogen bond to olefins and acetylenes is strongly catalyzed by platinum catalysts,[68] and this catalyzed addition reaction has been the subject of an intensive investigation by J. L. Speier and his co-workers.[69,70] In the presence of chloroplatinic acid, addition to terminal olefins proceeds readily at 100°. The addition to internal olefins, such as 2-pentene[69] or 3-heptene,[70] is much slower and yields only the n-pentyl- or n-heptylsilane derivative. Evidently, the olefin is first isomerized to the terminal position, and the latter reacts with the reagent.

$$n\text{-}C_2H_5CH{=}CHCH_3 \underset{(Pt)}{\rightleftharpoons} n\text{-}C_3H_7CH{=}CH_2$$
$$(Pt) \downarrow SiHCl_3$$
$$n\text{-}C_3H_7CH_2CH_2SiCl_3$$

In the case of cyclohexene, the double bond cannot migrate out of the ring, and the addition reaction to form the secondary derivative can be achieved.

[67] A. J. Barry, L. DePree, J. W. Gilkey, and D. E. Hook, J. Am. Chem. Soc., 69, 2916 (1947).

[68] G. H. Wagner and C. O. Strother, U.S. Patent 2,632,013, March 17, 1953.

[69] J. L. Speier, J. A. Webster, and G. H. Barnes, J. Am. Chem. Soc., 79, 974 (1957).

[70] J. C. Saam and J. L. Speier, J. Am. Chem. Soc., 80, 4104 (1958).

$$\text{cyclohexene} + SiH(CH_3)Cl_2 \xrightarrow[16 \text{ hr}, 100°]{H_2PtCl_6} \text{cyclohexyl-}Si(CH_3)Cl_2$$

It is interesting that the stereochemical results of the addition to acetylenes varies with the nature of the catalyst.[71]   Thus, the benzoyl peroxide catalyzed reaction results in a *trans* addition, whereas platinized charcoal gives *cis* addition.

$$RC\equiv CH + HSiCl_3 \xrightarrow{(RCOO)_2} \begin{array}{cc} R & SiCl_3 \\ | & | \\ C=C \\ | & | \\ H & H \end{array}$$

$$RC\equiv CH + HSiCl_3 \xrightarrow{(Pt)} \begin{array}{cc} R & H \\ | & | \\ C=C \\ | & | \\ H & SiCl_3 \end{array}$$

The addition of the aluminum-hydrogen bond to olefins, discovered[72] and explored[73] by Ziegler and his co-workers, is more closely related to the hydroboration reaction.

$$\begin{array}{c} | \ \ | \\ C=C \\ | \ \ | \end{array} + H—Al{\Huge\langle} \longrightarrow H—\begin{array}{c} | \\ C \\ | \end{array}—\begin{array}{c} | \\ C \\ | \end{array}—Al{\Huge\langle}$$

However, the reaction is much more sluggish.   Terminal olefins require temperatures in the neighborhood of 80 to 100° to achieve a modest rate of addition with either aluminum hydride or diethylaluminum hydride. For example, in adding diethylaluminum hydride to styrene, the reactants were heated to 100° for 17 hours.[74]

[71] R. A. Benkeser and R. A. Hickner, *J. Am. Chem. Soc.*, **80**, 5298 (1958).
[72] K. Ziegler, *Angew. Chem.*, **64**, 323, 330 (1952).
[73] K. Ziegler, Organo-Aluminum Compounds, in H. Zeiss (ed.), "Organometallic Chemistry," Chap. 5, American Chemical Society Monograph Series, Reinhold Publishing Corporation, New York, 1960.
[74] G. Natta, P. Pino, G. Mazzanti, P. Longi, and F. Bernardini, *J. Am. Chem. Soc.*, **81**, 2561 (1959).

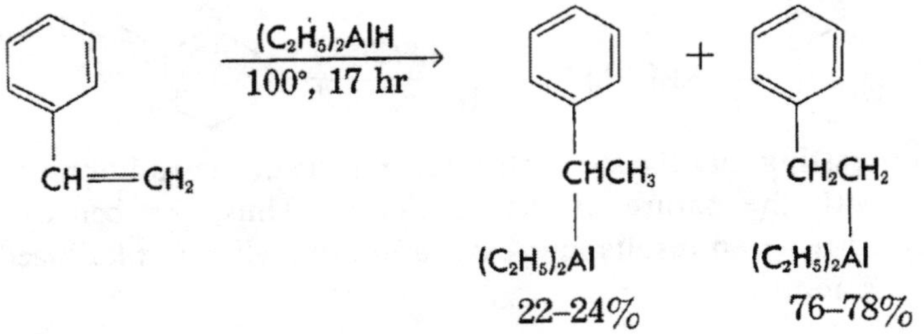

$$\text{(C}_2\text{H}_5)_2\text{AlH} \atop 100°, 17 \text{ hr}$$

CHCH₃ → $CHCH_3$

(C₂H₅)₂Al

22–24%

CH₂CH₂ → CH_2CH_2

(C₂H₅)₂Al

76–78%

The reaction with internal olefins is much more sluggish, and satisfactory addition has been achieved only with cyclic olefins, such as cyclopentene and cycloheptene, where the double bond is incapable of migrating to a terminal position.[75] Addition to acetylenes proceeds readily, and here the reaction evidently involves a simple *cis* addition of the aluminum-hydrogen linkage to the triple bond.[76]

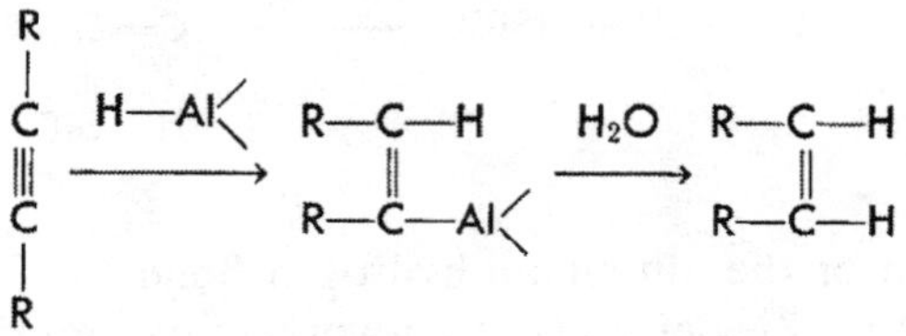

The organoaluminum compounds undergo isomerization and displacement reactions similar to those described for the organoboron derivatives.[73] Moreover, they undergo the important chain-growth reaction,[73] which has not yet been reported for the organoboranes. It is also important that they provide a possible entry into organoborane chemistry, through the reaction of triisobutylaluminum with methyl borate.

$$6(CH_3)_2C{=}CH_2 + 2Al + 3H_2 \longrightarrow 2[(CH_3)_2CHCH_2]_3Al$$

$$[(CH_3)_2CHCH_2]_3Al + (CH_3O)_3B \longrightarrow [(CH_3)_2CHCH_2]_3B + Al(OCH_3)_3$$

From triisobutylborane it is possible to synthesize many organoboranes via the displacement reaction.[22]

$$3RCH{=}CH_2 + [(CH_3)_2CHCH_2]_3B \longrightarrow (RCH_2CH_2)_3B + 3(CH_3)_2C{=}CH_2$$

[75] K. Ziegler, H. A. Gellert, H. Martin, K. Nagel, and J. Schneider, *Ann.*, **589**, 91 (1954).

[76] G. Wilke and H. Müller, *Ann.*, **618**, 267 (1958).

Finally, the addition of trialkyl- and triarylstannanes to a number of olefins at 80 to 100° has recently been achieved.[77–79] The addition also proceeds in the anti-Markownikoff direction.

Moreover, in contrast to the reactions involving aluminum hydride, many functional groups can be accommodated. Thus stannane derivatives have been added to acrylonitrile, methyl acrylate, and acrylamide.

$$(C_6H_5)_3SnH + CH_2{=}CHCN \longrightarrow (C_6H_5)_3SnCH_2CH_2CN$$

$$(n\text{-}C_3H_7)_3SnH + CH_2{=}CHCO_2CH_3 \longrightarrow (n\text{-}C_3H_7)_3SnCH_2CH_2CO_2CH_3$$

The addition of the silicon-hydrogen bond to olefinic and acetylenic derivatives constitutes a valuable synthetic route for the preparation of silicon derivatives. However, the latter are very stable substances, not readily transformed into other derivatives. Consequently, this reaction does not appear to be of interest as a general synthetic procedure.

The reactivity of the organoaluminum compounds is much more favorable for their utilization as intermediates in organic synthesis. However, their preparation via the addition of the aluminum-hydrogen bond

[77] G. J. M. van der Kerk, J. G. A. Luijten, and J. G. Noltes, *Chem. & Ind.* (*London*), **1956**, 352.

[78] G. J. M. van der Kerk, J. G. A. Luijten, and J. G. Noltes, *J. Appl. Chem.* (*London*), 7, 356 (1957).

[79] G. J. M. van der Kerk, J. G. A. Luijten, and J. G. Noltes, *Angew. Chem.*, **70**, 298 (1958).

to olefins is greatly restricted. First, the reaction proceeds really satis-factorily only with the lower 1-alkenes. Second, the reactivity of the aluminum-hydrogen bond is such that neither halogens nor any reducible functional groups can be tolerated. Third, the aluminum derivatives must compete in the laboratory with the more readily synthesized organolithium and organomagnesium compounds. Consequently, the organoaluminum compounds do not appear to have a promising future in laboratory syn-thetic work, although they will doubtless become of increasing importance in industrial chemistry because of their economical synthesis from alu-minum, hydrogen, and olefin.[73]

The new synthesis of organotin derivatives [77–79] could develop into a very useful synthetic procedure. Unfortunately, at the present time little is known about the applicability of these derivatives in organic synthesis. The utilization of this procedure suffers also at present from the disadvan-tage of requiring the prior synthesis of the organostannane intermediate.

At the present time, the wide generality of the hydroboration reaction, the ready synthesis of organoboranes from readily available and easily handled reagents, and the unique chemical properties of organoboranes all favor the development of this route as a most important synthetic tool for the organic chemist.

Conclusion

In the few short years since the hydroboration of olefins under mild conditions was first announced (1956),[1] extraordinary progress has been made in exploring this versatile reaction. It has not been possible within the scope of this survey to review all the literature and to present all the developments.

In the remaining chapters of this book, individual topics are considered in much greater detail. It is hoped that the more detailed treatment possible there will correct any injustices that may have been done as a result of the necessary selection of material to be presented in this intro-ductory survey.

2 | Early History

The first recorded observation of the reaction between an unsaturated hydrocarbon and diborane appears to be that of Alfred Stock, the father of diborane chemistry. He mentioned that acetylene reacts explosively with diborane at 100° to produce condensation products having an aromatic odor.[1]

No further mention of the reaction of diborane with unsaturated hydrocarbons appears until 1948. At that time Hurd reported that at elevated temperatures diborane reacts with olefins to form trialkylboranes, with benzene to form phenyl boron compounds, and with paraffins to form polymeric reaction products containing boron, carbon, and hydrogen.[2]

In a typical experiment ethylene containing 2 per cent by weight of diborane was heated under pressure in heavy-walled Pyrex tubes for 4 days at 100°. There was isolated a small amount of a clear liquid which inflamed on exposure to air. Mass spectrometric analysis indicated the presence of triethylborane.

Similarly large excesses of isobutylene were heated with diborane at 100° for 24 hours. Two fractions were obtained, b.p. 181.5° and 188.5°. These were identified as tri-*t*-butylborane and triisobutylborane, respectively. However, these identifications must now be considered questionable in view of the rapid isomerization exhibited by *t*-butylboron derivatives into isobutylboron structures.[3]

Hurd indicated that isobutylene also reacted with diborane at room temperature, but that the reaction proceeded even more slowly under these conditions.

[1] A. Stock and E. Kuss, *Ber.*, **56**, 789 (1923).

[2] D. T. Hurd, *J. Am. Chem. Soc.*, **70**, 2053 (1948).

[3] G. F. Hennion, P. A. McCusker, E. C. Ashby, and A. J. Rutkowski, *J. Am. Chem. Soc.*, **79**, 5190 (1957).

In 1950 Stone and Emeléus reported an interesting study of the capacity of diborane to function as a Lewis acid catalyst, in a manner similar to the related compound, boron trifluoride.[4] Indeed, they observed that at low temperatures diborane would cause the conversion of ethylene oxide and propylene oxide into polymers of a modest degree of polymerization. In following up this observation, they attempted to polymerize acrylonitrile, methyl methacrylate, and styrene. In each case the diborane was permitted to stand in contact with the olefin for 20 hours at room temperature. No polymerization of the olefins was observed. They indicated that a slow reaction occurred, with some formation of a trialkylborane, similar to that realized by Hurd at 100°. However, the complexity of the reaction products was discouraging. For example, they mention that they isolated at least three products from the reaction of acrylonitrile and diborane.

In the course of studying the combustion of hydrocarbons induced by aluminum borohydride, a slow reaction between 1-butene and the borohydride was observed.[5] This observation led to a study of the kinetics of the reaction of aluminum borohydride vapor with ethylene, propylene, and 1-butene.[5,6] The reaction was first order in aluminum borohydride and independent of the olefin concentration. At 140° the formation of triethylborane and an ethyl aluminum hydride derivative was noted.

$$\text{Al(BH}_4)_3 + 11\,C_2H_4 \xrightarrow{\;140°\;} 3(C_2H_5)_3B + \tfrac{1}{2}(C_2H_5)_4Al_2H_2$$

Finally, a kinetic study was made of the reaction of diborane with ethylene in the temperature range 120 to 175°.[7]

More recently, serious explosions were experienced in attempting to achieve the hydroboration of tetrafluoroethylene by heating the olefin with diborane at 80 to 100° for extended periods of time.[8]

Even under milder conditions the reactions of diborane with the four fluoroethylenes, tetrafluoro-, trifluoro-, 1,1-difluoro-, and monofluoroethylene, are difficult to control and yield complex mixtures.[9]

Diborane is a gas, unstable at moderate temperatures, with hazardous characteristics that have been well publicized. In view of these many

[4] F. G. A. Stone and H. J. Emeléus, *J. Chem. Soc.*, **1950**, 2755.

[5] R. S. Brokaw, E. J. Badin, and R. N. Pease, *J. Am. Chem. Soc.*, **72**, 1793 (1950).

[6] R. S. Brokaw and R. N. Pease, *J. Am. Chem. Soc.*, **72**, 5263 (1950).

[7] A. T. Whatley and R. N. Pease, *J. Am. Chem. Soc.*, **76**, 835 (1954).

[8] F. G. A. Stone and W. A. G. Graham, *Chem. & Ind.* (*London*), **1955**, 1181.

[9] B. Bartocha, W. A. G. Graham, and F. G. A. Stone, *J. Inorg. & Nuclear Chem.*, **6**, 119 (1958).

reports of the sluggishness and complexity of the reaction of diborane with olefins, this reaction did not appear to offer a promising preparative route for the synthesis of organoboranes. At the time we were engaged in exploring the factors influencing the reducing power of sodium borohydride. It is fortunate that an unexpected observation made in the course of this research program forced us to revise our ideas on the ease with which olefins could be converted into organoboranes and led us to embark on a program of exploration in this area.

It may be of interest to review briefly the background for the research program that led us so unexpectedly into this fascinating area.

During the years 1942–1943, in the course of war research at the University of Chicago, Professor H. I. Schlesinger and the present author together with our co-workers synthesized sodium borohydride for the first time and developed new practical syntheses for diborane, sodium borohydride, and other metal borohydrides[10] (Chapter 4). In the course of this research, it was observed that the alkali metal borohydrides were mild, but useful, reducing agents for aldehydes and ketones. Several years later, the related complex hydride, lithium aluminum hydride, was prepared and characterized. It proved to be a very powerful reducing agent.[11,12]

Unfortunately, the fact that the borohydride investigations had been carried out as part of a war research program caused it to be classified initially. For this and other reasons, the publication of this material was delayed for a number of years. The work with lithium aluminum hydride did not face these difficulties. Consequently, this product and its remarkable reducing potentialities were described to the chemical public considerably earlier than the corresponding data for the borohydrides.

Lithium aluminum hydride proved to be an exceedingly useful reagent for organic reductions. Numerous studies were devoted to exploring its full potentialities.[13] At the same time, relatively little attention was being devoted to sodium borohydride.

[10] H. I. Schlesinger and H. C. Brown, in collaboration with B. Abraham, A. C. Bond, N. Davidson, A. E. Finholt, J. R. Gilbreath, H. Hoekstra, L. Horvitz, E. K. Hyde, J. J. Katz, J. Knight, R. A. Lad, D. L. Mayfield, L. Rapp, D. M. Ritter, A. M. Schwartz, I. Sheft, L. D. Tuck, and A. O. Walker, *J. Am. Chem. Soc.*, **75**, 186 (1953).

[11] A. E. Finholt, A. C. Bond, Jr., and H. I. Schlesinger, *J. Am. Chem. Soc.*, **69**, 1199 (1947).

[12] R. F. Nystrom and W. G. Brown, *J. Am. Chem. Soc.*, **69**, 1197 (1947).

[13] W. G. Brown, Reductions by Lithium Aluminum Hydride, in R. Adams (ed.), "Organic Reactions," Vol. VI, Chap. 10, pp. 469–509, John Wiley & Sons, Inc., New York, 1951.

It was a considerable disappointment to the author, as one of the discoverers of sodium borohydride, to note that this promising reagent was being largely ignored in the rush to explore the applicability of the younger product. Accordingly, it was decided to institute a research program to explore the effects of solvents,[14] substituents,[15] and metal ions[14,16] on the reducing properties of borohydrides.

It was observed that the addition of aluminum chloride (or other polyvalent metal halides) to a solution of sodium borohydride in diglyme yielded a solution of greatly enhanced reducing power. At room temperature it rapidly reduced carboxylic acids, esters, and nitriles, groups that are normally resistant to the action of sodium borohydride alone.

The first exploration of the potentialities of the new reducing system was carried out conveniently by mixing the reagent in excess with the organic compound under study. An analysis of the reaction mixture for residual "hydride" indicated the extent of reaction.

Thus the treatment of nitriles with the reagent revealed the uptake of two "hydrides" per nitrile molecule. Evidently the reduction was proceeding to the amine stage. Similarly, esters such as ethyl benzoate and ethyl stearate required two hydrides per molecule, indicating reduction to the alcohol stage. However, the related ester, ethyl oleate, reacted with a disappearance of 2.4 to 2.5 hydrides per molecule. Purification of the ester did not alter the result. Use of a larger excess of reagent and a longer reaction time raised the uptake to 3.0 as a limit. Evidently the double bond was participating in the reaction.

Once this fact was recognized, it was a simple matter to demonstrate that olefins were capable of reacting quantitatively with the reagent to produce the corresponding organoboranes.[17,18]

The addition of aluminum chloride to a solution of sodium borohydride in diglyme in the stoichiometric quantity ($AlCl_3 + 3NaBH_4$) yields a clear colorless solution. The solution is quite stable with time. However, addition of an olefin, such as cyclopentene or 1-octene, to the solution results in the precipitation of a white solid as the reaction proceeds. Removal of the clear, supernatant liquid revealed that it contained all of the organically bound boron with only trace amounts of chloride ion and of residual

[14] H. C. Brown, E. J. Mead, and B. C. Subba Rao, J. Am. Chem. Soc., 77, 6209 (1955).

[15] H. C. Brown, E. J. Mead, and C. J. Shoaf, J. Am. Chem. Soc., 78, 3613 (1956).

[16] H. C. Brown and B. C. Subba Rao, J. Am. Chem. Soc., 78, 2582 (1956).

[17] H. C. Brown and B. C. Subba Rao, J. Am. Chem. Soc., 78, 5694 (1956).

[18] H. C. Brown and B. C. Subba Rao, J. Am. Chem. Soc., 81, 6423 (1959).

hydride. Analysis of the precipitate indicated the presence of aluminum and hydride in the ratio called for by aluminum hydride.

On this basis, the reaction proceeds in accordance with the equation

$$9RCH{=}CH_2 + 3NaBH_4 + AlCl_3 \longrightarrow 3(RCH_2CH_2)_3B + AlH_3 + 3NaCl$$

The reaction is relatively fast. Usually the reagent is treated with the olefin for 3 hours at room temperature and then heated briefly on the steam bath for $\frac{1}{2}$ hour to complete the reaction. In this way ethylene, propylene, 1-pentene, 2-pentene, 1-hexene, 1-octene, cyclopentene, cyclohexene, styrene, and α-methylstyrene are converted into the corresponding organoboranes in yields of 80 to 90 per cent of isolated products.

The observation that the solution of aluminum chloride and sodium borohydride is clear and does not precipitate sodium chloride, which is insoluble in diglyme, suggests that the reaction cannot proceed to the formation of aluminum borohydride. However, it is possible that a small equilibrium concentration of aluminum borohydride is formed and that the reaction proceeds through this intermediate. In this connection, it is of interest that aluminum borohydride in ether solution likewise serves to accomplish the hydroboration of olefins.

The application of this reagent for the hydroboration of olefins suffers from a major disadvantage. Only three of the four "hydrides" in the sodium borohydride molecule are utilized for the formation of the organoborane. The fourth is lost in the form of aluminum hydride. This difficulty was circumvented by the use of either boron trichloride or boron trifluoride in place of the aluminum chloride. With these two Lewis acids of boron, the reaction is amazingly fast — the olefin is converted into organoborane almost instantly at room temperature.[19]

$$12RCH{=}CH_2 + 3NaBH_4 + BCl_3 \longrightarrow 4(RCH_2CH_2)_3B + 3NaCl$$

$$12RCH{=}CH_2 + 3NaBH_4 + 4BF_3 \longrightarrow 4(RCH_2CH_2)_3B + 3NaBF_4$$

In the course of these studies it was observed that under the same experimental conditions, with ether solvents and at room temperature, diborane adds rapidly and quantitatively to olefins of widely differing structures.

$$6RCH{=}CH_2 + B_2H_6 \longrightarrow 2(RCH_2CH_2)_3B$$

[19] H. C. Brown and B. C. Subba Rao, *J. Am. Chem. Soc.*, **81**, 6428 (1959).

The reaction is rapid and quantitative. A number of attempts were made to measure the rate, but without success. The reaction appeared to be complete as fast as the two reagents could be brought into contact. In its speed, generality, and quantitative nature, the reaction resembles the addition of bromine to simple olefins.

These observations were exceedingly puzzling in view of the earlier reports. Accordingly we undertook a reexamination of the reaction of olefins with diborane in the high-vacuum apparatus. It was observed that the reaction of diborane with pure olefins is indeed quite slow, extending over periods of many hours. However, the reaction is subject to a marked catalysis by ethers and similar weak bases. Addition of mere traces of ethers changed the initially slow reaction to a fast one — too fast to measure by the usual manometric techniques.[20] Consequently, it was the inadvertent use of the ether solvents for the reactions involving sodium borohydride which appears to have been largely responsible for the fast, quantitative hydroboration procedures.

[20] H. C. Brown and L. Case, unpublished observations.

3 | Chemistry of Organoboranes

The first organoboranes were synthesized by Frankland in 1859, by the reaction of the newly discovered dialkylzinc compounds on triethyl borate.[1]

$$3(C_2H_5)_2Zn + 2(C_2H_5O)_3B \longrightarrow 2(C_2H_5)_3B + 3Zn(OC_2H_5)_2$$

He established that the products were stable to water and spontaneously inflammable in air. Frankland also observed the interesting ability of these compounds to combine with ammonia and other bases to form addition compounds.

$$(C_2H_5)_3B + NH_3 \longrightarrow (C_2H_5)_3B:NH_3$$

The organoboranes participate in Grignard-like reactions only sluggishly.[2,3] Moreover, their preparation involved the prior formation of a more reactive organometallic. As a result, there was little interest in applying these compounds to organic syntheses. In fact, until recently, most of the information available on the organoboranes were by-products of studies initiated primarily for theoretical considerations.

In this way the discovery that triarylboranes add sodium and other alkali metals to form addition compounds which are formally analogous to the triarylmethyl free radicals led to the synthesis of many triaryl-

[1] E. Frankland and B. F. Duppa, *Proc. Roy. Soc. (London)*, **10**, 568 (1859); *Ann.*, **115**, 319 (1860). E. Frankland, *J. Chem. Soc.*, **15**, 363 (1862); *Proc. Roy. Soc. (London)*, **12**, 123 (1863).

[2] H. Gilman and K. E. Marple, *Rec. trav. chim.*, **55**, 76, 133 (1936).

[3] H. Meerwein, G. Hinz, H. Majert, and H. Sönke, *J. prakt. Chem.*, **147**, 226 (1936).

boranes for a study of the influence of structure on the stability of these interesting alkali metal derivatives.[4-6]   Similarly, the reversible dissociation of addition compounds of organoboranes with amines provided a versatile tool for the study of steric effects.[7]   Finally, the similarity in the electronic structures of the trialkylboranes and carbonium ions led Johnson and his co-workers to make the first detailed study of the reactions of organoboranes.[8-11]

The rapid, quantitative conversion of clefins, dienes, and acetylenes into organoboranes, the main topic of this volume, is changing this situation.   There is a renewed interest in the organoboranes as intermediates for organic synthesis.[12]

It is not possible, within the space available, to present a complete, balanced, review of the chemistry of the organoboranes.   Consequently, this discussion will emphasize those aspects of the chemistry of aliphatic organoboranes of major interest because of their evident utility, in connection with hydroboration, to facilitate the work of the synthetic chemist. Reviews from a broader viewpoint are now available.[13,14]

Synthesis of Organoboranes via Organometallics

As was pointed out, the first synthesis of organoboranes utilized the reaction of dialkylzinc compounds on triethyl borate.[1]   With the discovery

[4] E. Krause, *Ber.*, **57**, 216 (1924).   E. Krause and H. Polack, *Ber.*, **59**, 777 (1926).   E. Krause and P. Nobbe, *Ber*, **63**, 934 (1930);   E. Krause and P. Nobbe, *Ber.*, **64**, 2112 (1931).

[5] H. E. Bent and M. Dorfman, *J. Am. Chem. Soc.*, **54**, 2132 (1932);   H. E. Bent and M. Dorfman, *J. Am. Chem. Soc.*, **57**, 1259, 1924 (1935).

[6] T. L. Chu, *J. Am. Chem. Soc.*, **75**, 1730 (1953).

[7] H. C. Brown, *J. Chem. Soc.*, **1956**, 1248.

[8] H. R. Snyder, J. A. Kuck, and J. R. Johnson, *J. Am. Chem. Soc.*, **60**, 105 (1938).

[9] J. R. Johnson, M. G. Van Campen, Jr., and O. Grummitt, *J. Am. Chem. Soc.*, **60**, 111 (1938).

[10] J. R. Johnson, H. R. Snyder, and M. G. Van Campen, Jr., *J. Am. Chem. Soc.*, **60**, 115 (1938).

[11] J. R. Johnson and M.G. Van Campen, Jr., *J. Am. Chem. Soc.*, **60**, 121 (1938).

[12] H. C. Brown, Organoboranes, in H. Zeiss (ed.), "Organometallic Chemistry," Chap. 4, American Chemical Society Monograph Series, Reinhold Publishing Corporation, New York, 1960.

[13] G. E. Coates, "Organo-Metallic Compounds," 2nd ed., Methuen & Co., Ltd., London, 1960, Chap. 3.

[14] M. F. Lappert, *Chem. Rev.*, **56**, 959 (1956).

and widespread utilization of the Grignard reagent, the latter became the preferred organometallic for this approach to the organoboranes. It was observed that boron trifluoride etherate provided better yields than the alkyl borates, so that the preferred synthetic route became the reaction of the Grignard reagent with boron trifluoride.[15]

With excess Grignard reagent the reaction frequently involves the formation of the tetraalkyl or tetraarylboron anion, BR_4^-. In the case of the alkyl derivatives, this offers no difficulty,[10] but the aryl derivatives are more stable and may require special methods for recovery of the organoborane.[16]

In the case of the t-butyl Grignard reagents, the synthesis is accompanied by a rearrangement from the t-butyl to the isobutyl structure.[17] Thus, the reaction of t-butylmagnesium chloride with boron trifluoride produces diisobutyl-t-butylborane.[18]

$$3(CH_3)_3CMgCl + BF_3 \longrightarrow (CH_3)_3CB[CH_2CH(CH_3)_2]_2 + 3MgClF$$

By taking advantage of this rearrangement it proved possible to synthesize an organoborane with three different alkyl groups attached to the boron atom.[19] The product, t-butylisobutyl-n-amylborane, proved to be surprisingly stable toward disproportionation, presumably a result of the large steric requirements of the alkyl substituents.

$$2(CH_3)_3CMgCl + n\text{-}C_5H_{11}BF_2 \longrightarrow n\text{-}C_5H_{11}\text{—}B \overset{\displaystyle C(CH_3)_3}{\underset{\displaystyle CH_2CH(CH_3)_2}{\big\langle}}$$

The reaction of the Grignard reagent with alkyl borates appears to be subject to more ready control, so that it has been widely utilized in devising synthetic routes to boronic and borinic esters and acids. In this way the aliphatic boronic esters are conveniently synthesized by the ad-

[15] E. Krause and R. Nitsche, *Ber.*, **54**, 2784 (1921).

[16] G. Wittig and P. Raff, *Ann.*, **573**, 195 (1951).

[17] G. F. Hennion, P. A. McCusker, E. C. Ashby, and A. J. Rutkowski, *J. Am. Chem. Soc.*, **79**, 5190 (1957).

[18] G. F. Hennion, P. A. McCusker, and A. J. Rutkowski, *J. Am. Chem. Soc.*, **80**, 617 (1958).

[19] G. F. Hennion, P. A. McCusker, and J. V. Marra, *J. Am. Chem. Soc.*, **80**, 3481 (1958).

dition of the Grignard reagent to the borate ester at low temperatures.[8,20] Similarly, the diarylborinic esters and acids have been obtained by the reaction of the aryl Grignard reagent with triisobutyl borate,[21] tri-n-butyl borate,[22] and trimethoxyboroxine.[23]

$$n\text{-}C_4H_9MgBr + B(OCH_3)_3 \xrightarrow{-78°} n\text{-}C_4H_9B(OCH_3)_2 + Mg(OCH_3)Br$$

$$2C_6H_5MgBr + B(OC_4H_9)_3 \longrightarrow (C_6H_5)_2BOC_4H_9 + 2Mg(OC_4H_9)Br$$

The recent developments in organoaluminum chemistry by Ziegler and his co-workers[24] have made these compounds readily available as industrial chemicals. Consequently, there has been renewed interest in the utilization of organoaluminum compounds for the synthesis of the corresponding organoboranes.[25,26]

It is possible to prepare trialkylboranes from the corresponding trialkylaluminum derivatives in good yield by utilizing either boron trifluoride[27] or borate esters.[28] Although the direct reaction of boric oxide with trialkylaluminum has not been successful, it has been possible to utilize the reaction product of trimethyl borate and boric oxide, namely, trimethoxyboroxine.[25,26]

$$R_3Al + BF_3 \longrightarrow R_3B + AlF_3$$

$$R_3Al + KBF_4 \longrightarrow R_3B + KAlF_4$$

$$R_3Al + B(OCH_3)_3 \longrightarrow R_3B + Al(OCH_3)_3$$

$$3R_3Al + (CH_3OBO)_3 \longrightarrow 3R_3B + Al(OCH_3)_3 + Al_2O_3$$

[20] P. A. McCusker and L. J. Glunz, *J. Am. Chem. Soc.*, **77**, 4253 (1955).

[21] N. N. Mel'nikov, *J. Gen. Chem. USSR*, **6**, 636 (1936); *C.A.*, **30**, 5571 (1936); N. N. Mel'nikov and M. S. Rokitskaya, *J. Gen. Chem. USSR*, **8**, 1768 (1938); *C.A.*, **33**, 4969 (1939).

[22] R. L. Letsinger and N. Remes, *J. Am. Chem. Soc.*, **77**, 2489 (1955).

[23] T. P. Povlock and W. T. Lippincott, *J. Am. Chem. Soc.*, **80**, 5409 (1958).

[24] K. Ziegler, Organo-Aluminum Compounds, in H. Zeiss (ed.), "Organometallic Chemistry," Chap. 5, American Chemical Society Monograph Series, Reinhold Publishing Corporation, New York, 1960.

[25] R. Köster, *Ann.*, **618**, 31 (1958).

[26] E. C. Ashby, *J. Am. Chem. Soc.*, **81**, 4791 (1959).

[27] E. Wiberg and K. Hertwig, cited by J. Goubeau, *FIAT Rev. Ger. Sci.*, **23**, 228 (1949).

[28] H. Gilman, PB Report 5596, OSRD No. 871, 1942.

Excellent yields have been realized in applying these methods to the synthesis of triethyl-, tri-*n*-propyl-, tri-*n*-butyl-, and triisobutylborane.[25,26] Although these methods should, in principle, be applicable to the synthesis of higher organoboranes, Köster indicates that the experimental difficulties increase with the higher members so that alternate syntheses become preferable.[25]

The aluminum alkyls also appear to provide a satisfactory route to the boronic esters by a carefully controlled reaction utilizing appropriate quantities of the reagents.[25,26]

$$(C_2H_5)_3Al + 3(n\text{-}C_4H_9O)_3B \longrightarrow 3C_2H_5B(O\text{-}n\text{-}C_4H_9)_2 + Al(O\text{-}n\text{-}C_4H_9)_3$$

In some cases the relatively inert organomercury compounds have been utilized for the synthesis of arylboron halides.[29]

$$(C_6H_5)_2Hg + BCl_3 \longrightarrow C_6H_5BCl_2 + C_6H_5HgCl$$

A new, promising procedure for the synthesis of aryldihaloboranes involves the reaction of boron trichloride or boron tribromide with the aromatic in the presence of aluminum powder.[30]

$$3C_6H_6 + 3BCl_3 + Al \longrightarrow 3C_6H_5BCl_2 + AlCl_3 + \tfrac{3}{2}H_2$$

The conditions are surprisingly mild: approximately 1 hour at 150° or 24 hours at 30 to 50° to produce yields of 60 to 70 per cent. In a related reaction aluminum bromide brings about a reaction of boron tribromide and benzene, providing phenyldibromoborane in 50 per cent yield.[31]

$$C_6H_6 + BBr_3 \xrightarrow[80°,\ 15\ hr]{AlBr_3} C_6H_5BBr_2 + HBr$$

[29] A. Michaelis, *Ber.*, **27**, 244 (1894); *Ann.*, **315**, 19, 26 (1901).
[30] E. L. Muetterties, *J. Am. Chem. Soc.*, **82**, 4163 (1960).
[31] Z. J. Bujwid, W. Gerrard, and M. F. Lappert, *Chem. & Ind. (London)*, **1959**, 1091.

Substitution reactions involving diborane and its derivatives at elevated temperatures also provide possible routes to the triarylboranes.[32,33]

$$3C_6H_6 + BH_3 \xrightarrow{\quad 100-180° \quad} (C_6H_5)_3B + 3H_2$$

Finally, related to the simple addition of the boron-hydrogen bond to olefins is the analogous addition of boron trihalides to certain activated olefins.[34]

Physical Properties

The aliphatic organoboranes are typical nonpolar substances resembling the corresponding hydrocarbons in their physical properties. Thus, trimethylborane is a gas, m.p. $-159.8°$, b.p. $-21.8°$, as compared to isobutane, m.p. $-159.6°$, b.p. $-11.7°$. Similarly, triethylborane exhibits the constants, m.p. $-92.5°$, b.p. $95°$, as contrasted with 3-ethylpentane, m.p. $-118.6°$, b.p. $93.5°$.

Alkylboranes containing long straight chains are low-melting solids but become liquids if the chains are branched. Naphthenic derivatives, such as tricyclopentylborane and tricyclohexylborane, tend to melt higher. Finally, the triarylboranes are generally crystalline solids, with relatively high melting points, e.g., triphenylborane, m.p. $142°$; tri-α-naphthylborane, m.p. 206 to $207°$.

The infrared spectra of the organoboranes reveal a band at 1125 cm^{-1} (8.16 μ) which has been assigned to the boron-carbon stretching vibration.[35]

Thermal Behavior

It was noted by Rosenblum that tri-n-butylborane undergoes decomposition slowly at temperatures of 100 to 130° to form n-butyldiboranes

[32] D. T. Hurd, *J. Am. Chem. Soc.*, **70**, 2053 (1948).

[33] R. Köster, K. Reinert, and K. H. Müller, *Angew. Chem.*, **72**, 78 (1960).

[34] M. F. Lappert, *Angew. Chem.*, **72**, 36 (1960).

[35] W. J. Lehman, C. O. Wilson, Jr., and I. Shapiro, *J. Chem. Phys.*, **28**, 777, 781 (1958).

and butenes.[36] Similarly, Köster has reported that in attempting to distill tri-n-decylborane he obtained 1-decene and n-decyldiboranes,[25] and Ashby has noted a smooth decomposition of tri-n-octylborane, upon attempted distillation at 169° at 0.2 mm, into *trans*-2-octene and tetra-n-octyldiborane.[26]

$$2(n\text{-}C_4H_9)_3B \xrightarrow{\Delta} 2n\text{-}C_4H_8 + [(n\text{-}C_4H_9)_2BH]_2$$

These results are all consistent in indicating a decomposition of the organoborane at moderate temperatures into the dialkylborane (dimer) and olefin. However, there appears to be some question as to the structure of the olefin produced in the reaction.

A somewhat different reaction was observed by Winternitz for tri-n-pentylborane.[37] Maintained at its boiling point, it evolved hydrogen and *trans*-2-pentene in equimolar amounts. The product was identified as a cyclic derivative, 1-n-pentyl-2-methylboracyclopentane.

$$(n\text{-}C_5H_{11})_3B \xrightarrow{\Delta} n\text{-}C_5H_{10} + H_2 +$$

It was also reported that tri-n-hexylborane reacts considerably slower, but likewise yields hydrogen, an olefin, and the related cyclic compound.

$$(n\text{-}C_6H_{13})_3B \xrightarrow{\Delta} n\text{-}C_6H_{12} + H_2 +$$

[36] L. Rosenblum, *J. Am. Chem. Soc.*, **77**, 5016 (1955).

[37] P. F. Winternitz and A. A. Carotti, *J. Am. Chem. Soc.*, **82**, 2430 (1960).

This reaction presumably proceeds through the following mechanism:

$$(n\text{-}C_5H_{11})_3B \xrightarrow{\Delta} (n\text{-}C_5H_{11})_2BH + n\text{-}C_5H_{10}$$

$$
\begin{array}{ccc}
H_2C\text{------}CH_2 & & H_2C\text{------}CH_2 \\
| \qquad\qquad | & \xrightarrow{\Delta} H_2 + & | \qquad\qquad | \\
H_2C \qquad CH_2\text{---}CH_3 & & H_2C \qquad CH\text{---}CH_3 \\
\searrow \quad H & & \searrow \quad \swarrow \\
\;\; B \;\;\nearrow & & B \\
| & & | \\
n\text{-}C_5H_{11} & & n\text{-}C_5H_{11}
\end{array}
$$

Attention should be called to one puzzling aspect of this experiment. Our own study of related systems indicates that primary carbon-hydrogen bonds participate more readily than secondary.[38]   This could account for the difference in rate between the n-pentyl and n-hexylboranes.   Saegebarth has reported that in the isomerization of cyclic organoboranes derived from dienes the reaction proceeds preferentially to the formation of a six-membered-ring derivative[39] (Chapter 15).   On this basis, the product from the cyclization of tri-n-pentylborane should be predominantly the 1-n-pentylboracyclohexane derivative.

$$
\begin{array}{c}
CH_2 \\
\diagup \quad \diagdown \\
H_2C \qquad\qquad CH_2 \\
| \qquad\qquad\qquad | \\
H_2C \qquad\qquad CH_2 \\
\diagdown \quad \diagup \\
B \\
| \\
n\text{-}C_5H_{11}
\end{array}
$$

In the hydroboration of *trans*-di-*t*-butylethylene, Logan and Flautt observed that the hydroboration ceases at the monoalkylborane stage. The product could be isolated, but lost hydrogen readily above 100° to form a cyclic organoborane.   Oxidation of the latter with alkaline hydrogen peroxide provided the glycol, 2,2,5,5-tetramethyl-1,4-hexanediol.[40]

[38] H. C. Brown, K. J. Murray, and G. Zweifel, unpublished observations.
[39] K. A. Saegebarth, *J. Am. Chem. Soc.*, **82**, 2081 (1960).
[40] T. J. Logan and T. J. Flautt, *J. Am. Chem. Soc.*, **82**, 3446 (1960).

$$\underset{\underset{H_3C}{\overset{H_3C}{\diagdown}}C\overset{CH_3}{\underset{CH_3}{\diagup}}}{\underset{H}{\overset{H}{\diagup}}}C=C\underset{H}{\overset{\overset{H_3C}{\diagup}\overset{CH_3}{\diagdown}}{\diagup}} \quad \xrightarrow{\text{HB}} \quad$$

$$\downarrow \Delta$$

$$\underset{\underset{HO}{\overset{CH_3}{|}}\underset{CH_3}{\overset{CH_3}{|}}}{H_2C-C-CH_2-CH-C-CH_3} \quad \xleftarrow{[O]} \quad + H_2$$

This reaction has been explored as a general synthesis of glycols from olefins.[38,41] Thus the hydroboration of 2,4,4-trimethyl-1-pentene to the dialkylborane stage, followed by a short period of heating at 200 to 220°, permits the synthesis of 2,4,4-trimethyl-1,5-pentanediol in 80 per cent yield, based on the equation below.[38]

$$2H_3C-\underset{\underset{CH_3}{|}}{\overset{\overset{CH_3}{|}}{C}}-CH_2-\underset{}{\overset{\overset{CH_3}{|}}{C}}=CH_2 \quad \xrightarrow{\text{HB}} \quad H_3C-\underset{\underset{CH_3}{|}}{\overset{\overset{CH_3}{|}}{C}}-CH_2-\underset{}{\overset{\overset{CH_3}{|}}{CH}}-CH_2$$

$$\downarrow 200°$$

$$\underset{\underset{OH}{|}\ \ \underset{CH_3}{|}}{CH_2-\overset{\overset{CH_3}{|}}{C}-CH_2-\underset{\underset{OH}{|}}{CH}-CH_2} \quad \xleftarrow{[O]} \quad + H_2$$

[41] R. Köster and G. Rotermund, *Angew. Chem.*, **72**, 138 (1960).

By heating tri-*n*-butylborane in an autoclave at 300°, Köster was able to achieve cyclization. Oxidation then afforded 1,4-butanediol in 80 per cent yield.[41]

$$(n\text{-}C_4H_9)_3B \xrightarrow{300°} \begin{array}{c} H_2C{-}{-}CH_2 \\ | \quad\quad | \\ H_2C \quad\quad CH_2 \\ \diagdown \quad \diagup \\ B \\ | \\ n\text{-}C_4H_9 \end{array} + H_2 + C_4H_8$$

$$\downarrow [O]$$

$$\underset{OH}{CH_2CH_2CH_2CH_2} + \underset{OH}{CH_3CH_2CH_2CH_2}$$

Because of the many possibilities for isomerization (Chapter 9), the products from higher boron alkyls, such as tri-(2-methyl-1-pentyl)borane, are mixtures and the reaction is less favorable for the synthesis of pure compounds.[38,41]

A much simpler, related reaction occurs with the aromatic derivatives.[42]

The noncyclic alkyl groups, R, can be removed by hydrogenation at 160° to provide products, such as boraindane, 3-methylboraindane, and bora-tetralin, dimers in solution.

[42] R. Köster and K. Reinert, *Angew. Chem.*, **71**, 521 (1959).

Finally, Köster reports that tri-*n*-octylborane yields bicyclic products in addition to the simple monocyclic derivatives of the type realized with tri-*n*-pentylborane and tri-*n*-hexylborane.[37] The conditions are rigorous — temperatures between 250 and 350° being employed.[43]

$$(n\text{-}C_8H_{17})_3B \longrightarrow \quad + \quad 2C_8H_{16} \quad + \quad 2H_2$$

$$(n\text{-}C_9H_{19})_3B \longrightarrow$$

CH₃ → CH_3

~8%   ~80%   ~12%

[O]

$$CH_2CH_2CH_2CH_2CHCH_2CH_2CH_2CH_2$$
$$\underset{OH}{|} \qquad \underset{OH}{|} \qquad \underset{OH}{|}$$

A very important reaction of the organoboranes is their isomerization at moderate temperatures. For example, it is reported that tri-*sec*-butylborane, prepared via the Grignard reaction, undergoes isomerization into tri-*n*-butylborane in the course of heating the material under reflux (200 to 215°) for 48 hours.[17] Under hydroboration conditions, this reaction is far faster (1 hour at 120 to 160°),[44,45] and it has proved to be of considerable importance. Consequently, discussion will be deferred until it is possible to discuss all pertinent aspects of the reaction (Chapter 9).

Behavior as Lewis Acids

The organoboranes possess a sextet of electrons and exhibit a strong tendency to share an electron-pair with a base, achieving thereby the more stable octet. Indeed, this strong tendency for organoboranes to react with such bases as ammonia and alkali metal hydroxides was recognized as early as 1862 by Frankland.[1]

$$(CH_3)_3B + NH_3 \longrightarrow (CH_3)_3B:NH_3$$
$$(CH_3)_3B + KOH \longrightarrow [(CH_3)_3B:OH]^-K^+$$

[43] R. Köster, *Angew. Chem.*, **72**, 563 (1960).
[44] H. C. Brown and B. C. Subba Rao, *J. Org. Chem.*, **22**, 1136 (1957).
[45] H. C. Brown and G. Zweifel, *J. Am. Chem. Soc.*, **82**, 1504 (1960).

In many of the reactions, to be discussed later in this chapter, the presence of a strong mineral base markedly enhances the desired chemical change. It is probable that the coordination of the hydroxide ion with the boron alkyl greatly increases the carbanion-like character of the alkyl groups and thereby facilitates reactions in which the alkyl group is transferred with its electron pair.

$$R\text{—}B\text{—}R + OH^- \longrightarrow \left[R\overset{\delta^-}{\text{—}}B\text{—}OH \right]$$

Trimethylborane readily adds in a 1 : 1 ratio to lithium hydride suspended in ether. In this case the reaction is reversible, the product dissociating into its components at temperatures above 100°.

$$LiH + B(CH_3)_3 \rightleftharpoons Li^+[H:B(CH_3)_3]$$

The corresponding derivatives of sodium hydride are more stable, exhibiting no tendency toward dissociation.[46] Similar derivatives of alkali metal hydrides and triarylboranes have been described.[47]

It was observed in 1940 that ethyllithium reacts readily with trimethylborane to form the quaternary boron derivative, $Li^+[C_2H_5B(CH_3)_3]^-$.[48] Many related products are now known, including lithium tetramethylboron, $Li^+[(CH_3)_4B]^-$,[49] and sodium tetraphenylboron, $Na^+[(C_6H_5)_4B]^-$.[47] The latter derivative is quite stable in aqueous solution and has found considerable application as a precipitating agent in analytical chemistry.

The study of the addition compounds of organoboranes with amines has provided a valuable tool in recent years for the study of steric effects.[7] For example, a simple experiment which demonstrates the importance of steric effects in these addition compounds involves a comparison of the relative stabilities of the addition compounds of ammonia and of trimethylamine with trimethylborane and triisopropylborane. Thus the addition compound of trimethylborane with trimethylamine is more stable

[46] H. C. Brown, H. I. Schlesinger, I. Sheft, and D. M. Ritter, *J. Am. Chem. Soc.*, **75**, 192 (1953).

[47] G. Wittig and A. Rückert, *Ann.*, **566**, 101 (1950).

[48] H. I. Schlesinger and H. C. Brown, *J. Am. Chem. Soc.*, **62**, 3429 (1940).

[49] D. T. Hurd, *J. Org. Chem.*, **13**, 711 (1948).

than the corresponding product from ammonia. This increased stability is attributed to the inductive effect of the three methyl groups in trimethyl-amine which increases the electron density on the nitrogen atom and thereby renders it better able to share its electron pair with the boron atom. However, with triisopropylborane as the reference acid, ammonia forms a far more stable addition compound than trimethylamine. This reversal is attributed to the effect of the large steric interactions between the isopropyl groups of the borane and the methyl groups of the amine.[50]

$$Me_3B:NH_3 + NMe_3 \rightleftharpoons Me_3B:NMe_3 + NH_3$$

$$i\text{-}Pr_3B:NH_3 + NMe_3 \rightleftharpoons i\text{-}Pr_3B:NMe_3 + NH_3$$

These addition compounds dissociate reversibly in the gas phase. An examination of the change with temperature of the equilibrium constant for the dissociation leads to precise values for $\Delta F°$, $\Delta H°$, and $\Delta S°$ of dissociation.[51]

The heat of dissociation of trimethylamine-trimethylborane is 17.6 kcal/mole, whereas that for triethylamine-trimethylborane is much less, in the neighborhood of 10 kcal/mole. This difference in stability does not appear explicable in terms of the great similarity in the electrical effects of methyl and ethyl groups. It is explicable in terms of the large difference in the steric requirements of the two bases.

This conclusion is confirmed by an examination of the behavior of quinuclidine, which may be considered to be related to triethylamine with the three ethyl groups tied back so they cannot interfere with the tri-

$$\Delta H = -17.6$$

$$\Delta H = -10$$

[50] H. C. Brown, *J. Am. Chem. Soc.*, **67**, 374 (1945).

[51] H. C. Brown, M. D. Taylor, and M. Gerstein, *J. Am. Chem. Soc.*, **66**, 431 (1944).

methylborane molecule.   Here we observe an energy of dissociation of 20.0 kcal/mole.[52]

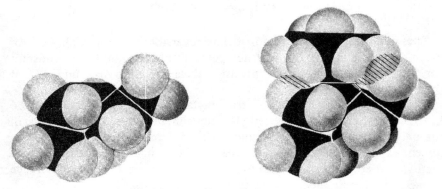

$$\Delta H = -20.0$$

The arguments may be clarified by examining Figures 3–1 and 3–2.

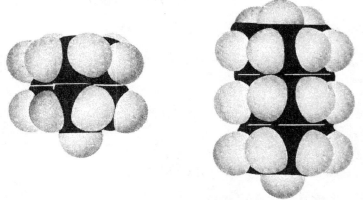

Figure 3-1   Molecular models of triethylamine and triethylamine-trimethylborane.   Cross-hatched area shows conflicting steric requirements.

Figure 3-2   Molecular models of quinuclidine and quinuclidine-trimethylborane.

[52] H. C. Brown and S. Sujishi, *J. Am. Chem. Soc.*, **70**, 2878 (1948).

Such studies have contributed greatly to placing the study of steric effects on a quantitative basis.[7,53]

Triarylboranes likewise react readily with amine bases and numerous such compounds have been synthesized.[54] Unfortunately, no quantitative data are presently available on the stabilities of these compounds.

Redistribution Reactions

It was observed by Schlesinger and his co-workers that diborane reacts readily with trimethyl-,[55] triethyl-,[56] and tri-n-propylborane[56] to yield a mixture of the corresponding alkyldiboranes.

$$R_3B + B_2H_6 \longrightarrow \quad RHBH_2BH_2$$
$$+ R_2BH_2BH_2$$
$$+ R_2BH_2BHR$$
$$+ R_2BH_2BR_2$$

The sym-dimethyldiborane, $CH_3HBH_2BHCH_3$, was prepared later by the reaction of monomethyldiborane with dimethyl ether.[57] The latter forms an addition compound with borane, but not with methylborane.

$$CH_3HBH_2BH_2 + (CH_3)_2O \xrightarrow{-80°} CH_3BH_2 + (CH_3)_2O : BH_3$$

$$2CH_3BH_2 \longrightarrow CH_3HBH_2BHCH_3$$

Recently, sym-diethyldiborane was isolated from the disproportionation of ethyldiborane.[58] The product proved to be quite stable to disproportionation.

$$2C_2H_5HBH_2BH_2 \longrightarrow C_2H_5HBH_2BHC_2H_5 + B_2H_6$$

[53] H. C. Brown, *J. Chem. Educ.*, **36**, 424 (1959).

[54] E. Krause and A. von Grosse, "Die Chemie der metallorganischen Verbindungen," Borntraeger, Berlin, 1937.

[55] H. I. Schlesinger and A. O. Walker, *J. Am. Chem. Soc.*, **57**, 621 (1935).

[56] H. I. Schlesinger, L. Horvitz, and A. B. Burg, *J. Am. Chem. Soc.*, **58**, 407 (1936).

[57] H. I. Schlesinger, N. W. Flodin, and A. B. Burg, *J. Am. Chem. Soc.*, **61**, 1078 (1939).

[58] I. J. Solomon, M. J. Klein, and K. Hattori, *J. Am. Chem. Soc.*, **80**, 4520 (1958).

The trialkylboranes are monomers.   However, all derivatives from BH_3 to RBH_2 and R_2BH appear to exist in the dimeric form with a double hydrogen bridge.   The alkyl-boron linkage is quite stable.   Consequently, the rapid exchange of alkyl groups and hydrogen in the redistribution reaction must involve an intermediate in which the shift can occur without any great energy requirement.   It would appear adequate to postulate that a mixture of diborane and trialkylborane will contain a small quantity of an unsymmetrical bridged species, which can break down either into its components, or into the alkylboranes.

Combinations of BH_3, RBH_2, and R_2BH with each other will lead to all the observed products.

There is accumulating evidence of another equilibrium which does not involve transfer of the alkyl groups from the boron atom to which they are attached.   This equilibrium is fast, occurring rapidly even at $0°$.[58,59]

$$(RBH_2)_2 + (BH_3)_2 \rightleftharpoons 2RHBH_2BH_2$$

$$(R_2BH)_2 + (BH_3)_2 \rightleftharpoons 2R_2BH_2BH_2$$

This subject is discussed further in Chapter 12.

It therefore appears that equilibration of trialkylboranes with excess diborane provides a convenient route to the monoalkylboranes (or their dimers).   Since these monoalkylboranes undergo ready alcoholysis to the corresponding boronic esters, or hydrolysis to the boronic acids, hydroboration provides a convenient route to these derivatives.[60]

[59] H. C. Brown and G. J. Klender, *J. Inorg. Chem.*, in press.
[60] H. C. Brown, A. Tsukamoto, and D. B. Bigley, *J. Am. Chem. Soc.*, **82**, 4703 (1960).

$$3n\text{-}C_3H_7CH\!=\!CH_2 \xrightarrow[\text{THF}]{\text{BH}_3} (n\text{-}C_5H_{11})_3B$$

$$\downarrow \text{BH}_3$$

$$3n\text{-}C_5H_{11}B(OCH_3)_2 \xleftarrow{\text{CH}_3OH} 3n\text{-}C_5H_{11}BH_2$$

By controlling the amount of excess diborane, it is possible to direct the equilibration reaction to the dialkylborane stage. This provides a convenient route to the dialkylborinic acids and esters.[60]

The redistribution reaction between diborane and triphenylborane has been utilized for the synthesis of *sym*-diphenyldiborane.[61]

$$2(C_6H_5)_3B + 2B_2H_6 \xrightarrow[\text{2.2 atm}]{80°} 3C_6H_5HBH_2BHC_6H_5$$

It is of considerable interest that triethyl- and tri-*n*-propylborane fail to undergo exchange below 100° in the absence of a catalyst. However, in the presence of minor amounts of triethylaluminum or diborane, rapid exchange occurs to give the statistically redistributed products, $B(C_2H_5)_3$, $n\text{-}C_3H_7B(C_2H_5)_2$, $(n\text{-}C_3H_7)_2BC_2H_5$, $(n\text{-}C_3H_7)_3B$.[62] Presumably, the triethylaluminum, which can form dimers involving alkyl bridges, must form similar bridges with the boron, albeit in low concentrations, and thereby facilitate an exchange of groups.

$$R_3B + R'_3Al \rightleftharpoons R'_2Al\!\!\begin{array}{c} R' \\ \diagup \diagdown \\ \diagdown \diagup \\ R' \end{array}\!\!BR_2 \rightleftharpoons R'_2AlR + R_2BR'$$

Added catalysts are not required to achieve an exchange between boron trihalides and trialkylboranes. For example, boron trichloride reacts rapidly with trialkylboranes at temperatures above 100° to give nearly quantitative yields of dialkylchloroboranes.[63,64] The reaction has

[61] E. Wiberg, J. E. F. Evans, and H. Nöth, *Z. Naturforsch.*, **13b**, 263 (1958).

[62] R. Köster and G. Bruno, *Ann.*, **629**, 89 (1960).

[63] V. W. Buls, O. L. Davis, and R. I. Thomas, *J. Am. Chem. Soc.*, **79**, 337 (1957).

[64] P. A. McCusker, G. F. Hennion, and E. C. Ashby, *J. Am. Chem. Soc.*, **79**, 5192 (1957).

been demonstrated for tri-*n*-butylborane, triisobutylborane, and tri-*sec*-butylborane, so that it is relatively insensitive to the structure of the alkyl group.

The dialkylchloroboranes may be distilled without change at temperatures under 100°. At elevated temperatures an equilibrium is set up, $2R_2BCl \rightleftharpoons RBCl_2 + R_3B$, from which the more volatile monoalkyldichloroborane may be fractionated. Alternately, the monoalkyldichloroborane or the monoalkyldifluoroborane may be obtained by heating the trialkylborane with the boron halide in an autoclave at 200° for 20 hours.[63]

At 200° it is possible to achieve a slow reversible reaction between boric oxide and trialkylboranes, forming the corresponding trialkylboroxine.[65]

In a typical experiment tri-*n*-butylborane and boric oxide, in equimolar amounts, were heated under reflux for 40 hours, and the product, tri-*n*-butylboroxine, was recovered by vacuum distillation of the reaction mixture. The yield realized was approximately 70 per cent. The reaction appears to be limited to organoboranes, whose groups are not susceptible to the isomerization reaction. Thus, triisobutyl-, tricyclohexyl-, and triphenylborane are all converted satisfactorily into the corresponding boroxines. However, tri-*sec*-butylborane was converted into tri-*n*-butylboroxine.

Finally, Köster has reported, without experimental details, that it is possible to catalyze with small quantities of diborane a redistribution reaction at low temperatures between trialkylboranes and borate esters or boron halides.[66]

Protonolysis

The trialkylboranes are remarkably stable toward water. Ulmschneider and Goubeau report that the hydrolysis of trimethylborane with an

[65] G. F. Hennion, P. A. McCusker, E. C. Ashby, and A. J. Rutkowski, *J. Am. Chem. Soc.*, **79**, 5194 (1957).

[66] R. Köster, *Angew. Chem.*, **73**, 66 (1961).

equivalent of water for 7 hours at 180° affords a 69 per cent yield of dimethylborinic acid.[67] The reaction with hydrogen sulfide is even more sluggish — a temperature of 280° is required to achieve conversion to boron sulfide.[67]

The reactions with alcohols and phenols are also slow. Very high temperatures were required to achieve the indicated reaction with ethylene glycol and catechol.[67]

The addition of alkali to water stabilizes the trialkylboron toward hydrolysis.[68] Treatment with concentrated mineral acids facilitates the hydrolysis. However, even heating under reflux with such acids brings about the loss of only one alkyl group. Thus, Johnson et al. report that in heating tri-n-butylborane with 48 per cent hydrobromic acid under reflux a quantitative yield of di-n-butylborinic acid was obtained in 1 hour.[10]

$$(n\text{-}C_4H_9)_3B + HBr \longrightarrow (n\text{-}C_4H_9)_2BBr + n\text{-}C_4H_{10}$$

$$(n\text{-}C_4H_9)_2BBr + H_2O \longrightarrow (n\text{-}C_4H_9)_2BOH + HBr$$

The cessation of reaction at this stage is not due to the rapid hydrolysis of the presumed intermediate, di-n-butylboron bromide, since this compound is readily produced by treatment of tri-n-butylborane with an excess of anhydrous hydrogen bromide at 55 to 60°, but further dealkylation is not observed unless much more vigorous conditions are employed.

Somewhat unexpectedly, the organoboranes have proved to be more susceptible to attack by carboxylic acids than to mineral acids. Thus, it

[67] D. Ulmschneider and J. Goubeau, *Ber.*, **90**, 2733 (1957).
[68] H. C. Brown and N. C. Hébert, unpublished observations.

has been observed that under relatively mild conditions triethylborane can be converted into diethylboron acetate and ethane,[69,70]

$$(C_2H_5)_3B + CH_3CO_2H \xrightarrow{100°} (C_2H_5)_2BOCOCH_3 + C_2H_6$$

A detailed study of the action of carboxylic acids on organoboranes has revealed that two of the three groups can be removed by excess anhydrous acid at room temperature, and all three groups can generally be removed by refluxing the organoborane in diglyme solution with a moderate excess of propionic acid for 2 to 3 hours.[71] Consequently, hydroboration of olefins in diglyme, followed by refluxing with propionic acid, offers a convenient noncatalytic procedure for the hydrogenation of double bonds.

$$3RCH{=}CH_2 \xrightarrow{HB} 3(RCH_2CH_2)_3B \xrightarrow{C_2H_5CO_2H} 3RCH_2CH_3$$

Secondary alkyl groups undergo protonolysis less readily than primary. Consequently, in hydrogenating internal olefins, especially those containing bulky groups, it is preferable that the boron atom be transferred to the terminal position by heating under reflux (Chapter 9) prior to the addition of the acid. Moreover, use of the less volatile solvent, triglyme, permits completion of the protonation stage in $\frac{1}{2}$ to 1 hour.

The combined hydroboration-protonolysis procedure has been applied to a considerable number of representative olefins, such as 1-hexene, 2-hexene, 1-octene, 2,4,4-trimethyl-1-pentene, 2,4,4-trimethyl-2-pentene, cyclopentene, and cyclohexene. Yields of 80 to 90 per cent of the hydrogenated product are realized.

Since olefins containing active sulfur, chlorine, and nitrogen substituents readily undergo hydroboration, this procedure opens up the possibility of hydrogenating olefinic derivatives containing such groups. In this way allylmethylsulfide is converted into n-propylmethylsulfide in a yield of 78 per cent.

[69] H. Meerwein, G. Hinz, H. Majert, and H. Sönke, *J. prakt. Chem.*, **147**, 226 (1936).

[70] J. Goubeau, R. Epple, D. Ulmschneider, and H. Lehmann, *Angew. Chem.*, **67**, 710 (1955).

[71] H. C. Brown and K. Murray, *J. Am. Chem. Soc.*, **81**, 4108 (1959).

The reaction appears to proceed with retention of configuration. Thus tri-*exo*-norbornylborane undergoes deuterolysis to yield *exo*-deuteronorbornane.[72]

The stereochemistry of this and related reactions is discussed in more detail in Chapter 8.

The complete protonolysis of the alkylboranes has been applied for the quantitative analysis of these derivatives.[73]

Organoboranes of the vinyl type, readily obtained by the hydroboration of acetylenes (Chapter 16), undergo protonolysis much more readily. These compounds undergo complete protonolysis with acetic acid at 0°, and the reaction provides a convenient means of converting acetylenes into *cis* olefins of high purity.[74]

Here also, it has been demonstrated that carboxylic acids are especially effective. The protonolysis fails to proceed under these conditions with strong mineral acids, such as hydrochloric or sulfuric acid.[75]

It would appear that the unique effectiveness of carboxylic acids must derive from the structure of the carboxylic acid group. Presumably the

[72] H. C. Brown and K. J. Murray, *J. Org. Chem.*, **26**, 631 (1961).

[73] J. Crighton, A. K. Holliday, A. G. Massey, and N. R. Thompson, *Chem. & Ind. (London)*, **1960**, 347.

[74] H. C. Brown and G. Zweifel, *J. Am. Chem. Soc.*, **81**, 1512 (1959).

[75] H. C. Brown and D. Bowman, unpublished observations.

trialkylborane is capable of coordinating at the oxygen of the carbonyl group, placing the boron-carbon bond in position for an intramolecular attack by the proton.

The increasing difficulty in each successive stage presumably arises in large part from the decreased ability of the dialkylboron acetate and the monoalkylboron diacetate to accept an electron pair and become coordinated to the carbonyl group.

Hydrogenation

At elevated temperatures (140 to 160°) and high pressures of hydrogen (200 to 300 atm) organoboranes undergo partial hydrogenation to alkanes and alkylated diboranes. Unfortunately, few details have yet appeared in the scientific literature,[76] but it appears that the reaction is relatively sluggish[77] and not convenient as a means of converting the organoborane into the corresponding alkane.

In the presence of a tertiary amine, it is possible to operate at higher temperatures, 200 to 220°. In this way, in the presence of triethylamine, triethylborane was converted into triethylamine-borane and ethane in essentially quantitative yield.[78]

$$(C_2H_5)_3N + B(C_2H_5)_3 + 3H_2 \xrightarrow[\text{300 atm}]{200-220°} (C_2H_5)_3N:BH_3 + 3C_2H_6$$

[76] R. Köster, Angew. Chem., 68, 383 (1956).
[77] H. E. Podall, H. E. Petree, and J. R. Zietz, J. Org. Chem., 24, 1222 (1959).
[78] R. Köster, Angew. Chem., 69, 94 (1957).

Halogenation

The aliphatic organoboranes exhibit a peculiar inertness toward the halogens. For example, tri-*n*-butylborane reacts slowly with bromine to form di-*n*-butylboron bromide and *n*-butyl bromide.[10]

$$(n\text{-}C_4H_9)_3B + Br_2 \longrightarrow (n\text{-}C_4H_9)_2BBr + n\text{-}C_4H_9Br$$

However, reaction beyond this point is quite difficult. If an attempt is made to force the reaction, the product shows evidence of substitution in the alkyl groups.

Similarly, it is reported that a temperature of 150° is conducive for the reaction of iodine with tri-*n*-propylborane to replace one of the three alkyl groups.[79]

Oxidation — Oxygen

The lower alkylboranes are spontaneously inflammable in air and burn with a green flame. Tri-*n*-butylborane and the higher members retain their sensitivity to oxygen but do not inflame. The triarylboranes are also sensitive to oxygen, but derivatives containing large bulky groups, such as tri-α-naphthylborane[80,81] and trimesitylborane,[82] are relatively stable to oxygen and can be handled in air without special precautions.

Oxidation at room temperature of the higher aliphatic boranes with moist air results in the uptake of one equivalent of oxygen and the formation of the corresponding alkylborinate, R_2BOR. However, under anhydrous conditions, an alkylboronate, $RB(OR_2)$, is formed.[11]

$$R_3B + \tfrac{1}{2}O_2 \xrightarrow{\text{moist air}} R_2BOR$$

$$R_3B + O_2 \xrightarrow{\text{dry air}} RB(OR)_2$$

The inhibiting effect of the moisture on the further oxidation of the borinate has been attributed to the formation of a reasonably stable hydrate by the alkylborinate, which prevents further coordination by

[79] L. H. Long and D. Dollimore, *J. Chem. Soc.*, **1953**, 3902, 3906.

[80] E. Krause and P. Nobbe, *Ber.*, **63**, 934 (1930).

[81] H. C. Brown and S. Sujishi, *J. Am. Chem. Soc.*, **70**, 2793 (1948).

[82] H. C. Brown and V. H. Dodson, *J. Am. Chem. Soc.*, **79**, 2302 (1957).

oxygen. Johnson and Van Campen proposed that the oxidation of the organoborane proceeds through an initial addition compound of the trialkylborane and oxygen, $R_3B \cdot O_2$, which they term a "borine peroxide," followed by reaction of this intermediate with a second mole of the trialkylborane.

$$R_3B + O_2 \longrightarrow R_3B \cdot O_2$$

$$R_3B \cdot O_2 + R_3B \longrightarrow 2R_2BOR$$

The borinate can react further, unless its coordination with oxygen is prevented by water.

Detailed studies of the oxidation of the lower alkylboranes have refined this early interpretation. Thus Petry and Verhoek report that in a flow system (room temperature, 10 to 15 mm of pressure, contact time of 2 to 3 minutes) trimethylborane reacts with oxygen quantitatively to form dimethylboron methylperoxide.[83]

$$(CH_3)_3B + O_2 \longrightarrow (CH_3)_2BOOCH_3$$

Similarly, Abraham and Davies observed that in dilute cyclohexane solution, tri-n-butylborane absorbed 1.2 moles of oxygen, giving a product that contained 1.06 moles of peroxide (by iodometric titration). Under the same conditions, oxidation of tri-t-butylborane(?)[17] yielded a peroxide with the formula t-BuB(OOt-Bu)$_2$. t-Butylhydroperoxide was identified as a product of alkaline hydrolysis of this intermediate.[84]

It has been suggested by Davies and Moodie that the autooxidation of trialkylboranes proceeds initially by the coordination of the oxygen molecule to the boron atom, followed by a migration of alkyl groups from boron to oxygen.[85]

$$R_3B + O_2 \longrightarrow R{-}\overset{\displaystyle R}{\underset{\displaystyle R}{\overset{|}{\underset{|}{B^-}}}}{-}O{-}O^+ \longrightarrow R{-}\overset{\displaystyle R}{\underset{\displaystyle R}{\overset{|}{\underset{|}{B}}}}{-}OOR \qquad \text{etc.}$$

[83] R. C. Petry and F. H. Verhoek, *J. Am. Chem. Soc.*, **78**, 6416 (1956).
[84] M. H. Abraham and A. G. Davies, *Chem. & Ind.* (*London*), **1957**, 1622.
[85] A. G. Davies and R. B. Moodie, *Chem. & Ind.* (*London*), **1957**, 1622.

It has been reported that the trialkylboranes are effective catalysts for vinyl polymerization.[86] However, it now appears that it is the alkylboron peroxides, resulting from the presence of minor amounts of oxygen, that serve as the active catalyst.[87]

On the basis of earlier work[11] it had been considered that only two of the three alkyl groups could be conveniently converted by air oxidation into the alcohol. However, it has been demonstrated that with the proper control of the experimental conditions it is possible to achieve complete conversion of all three groups.[88]

Oxidation — Alkaline Hydrogen Peroxide

It was early demonstrated by Johnson and Van Campen[11] that perbenzoic acid in chloroform solution at 0° reacts quantitatively with tri-n-butylborane, and the reaction can be utilized for the quantitative analysis of organoboranes. A reaction of benzoyl peroxide was observed also, but the results proved to be erratic. In this paper it is stated that aqueous hydrogen peroxide, in the presence of dilute alkalies, effects a complete dealkylation of tri-n-butylborane with the formation of boric acid and n-butyl alcohol. The reaction was suggested as the basis of a convenient method for the determination of boron in organoboron compounds. Unfortunately, no experimental details were reported.

The reaction was developed as an analytical procedure by Belcher, Gibbons, and Sykes in 1952.[89] Their conditions were exceedingly vigorous. They recommended treating the organoborane with excess hydrogen peroxide and concentrated sodium hydroxide, under reflux. Essentially these conditions were utilized by Winternitz and Carotti[37] in their study of the boron heterocycles formed in the thermal decomposition of tri-n-pentylborane and tri-n-hexylborane. The major products realized in their oxidation were the cyclic ethers, 2-methyltetrahydrofuran and 2-methyltetrahydropyran, and not the simple glycols.

Our own early applications of this method were also influenced by the

[86] G. S. Kolesnikov and N. V. Klimentova, *Izvest. Akad. Nauk SSSR*, **1957**, 652; *C. A.*, **51**, 15458 (1957). G. S. Kolesnikov and L. S. Fedorova, *Izvest. Akad. Nauk SSSR*, **1957**, 236; *C.A.*, **51**, 11291 (1957).

[87] N. Ashikari, *J. Polymer Sci.*, **28**, 250 (1958); N. Ashikari, *Bull. Chem. Soc. Japan*, **31**, 229 (1958); J. Furukawa and T. Tsuruta, *J. Polymer Sci.*, **28**, 227 (1958).

[88] S. B. Mirviss, *J. Am. Chem. Soc.*, **83**, 3051 (1961).

[89] R. Belcher, D. Gibbons, and A. Sykes, *Mikrochim. Acta*, **40**, 76 (1952).

vigorous conditions recommended for the analytical procedure.[90,91] However, with continued use of this reaction it was observed that the milder the conditions, the better the yield. In view of the growing importance of this reaction in synthetic chemistry,[92] it appeared desirable to undertake a detailed study of the reaction of alkaline hydrogen peroxide with trialkylboranes.[93]

$$R_3B + 3H_2O_2 + NaOH \longrightarrow 3ROH + NaB(OH)_4$$

It was observed that the treatment of the organoborane from the usual hydroboration of 50 mmoles of 1-hexene in 40 ml of diglyme with 15 mmoles of sodium hydroxide (5 ml of a 3 M solution), followed by the slow addition (rapid vigorous reaction!) of 60 mmoles of hydrogen peroxide, 20 per cent excess (6.0 ml of 30 per cent hydrogen peroxide) at 25 to 30° led to essentially instantaneous oxidation, with yields of 94 to 97 per cent of alcohol based on the original olefin utilized. The reaction proceeded satisfactorily even at 0°, with a yield of 89 per cent realized.

The three solvents commonly utilized for hydroborations are diglyme, tetrahydrofuran, and ethyl ether. It was observed that the oxidation proceeded with equal ease in tetrahydrofuran. However, the reaction in ethyl ether was more sluggish. Both diglyme and tetrahydrofuran are partially miscible with the aqueous phase in these oxidations, whereas ethyl ether is not. However, the simple addition of ethanol as a co-solvent avoided the difficulty in the ethyl ether system and brought about an increase in the yield to 98 per cent.

Wide variations in the structure of the olefin can be tolerated in this reaction. The oxidation appears to be remarkably free of the steric requirements of the alkyl group in the organoborane, as indicated by the essentially quantitative conversions realized with trinorbornylborane, diisopinocampheylborane, and similar organoboranes containing large bulky alkyl groups.

[90] H. C. Brown and B. C. Subba Rao, *J. Am. Chem. Soc.*, **78**, 5694 (1956).

[91] H. C. Brown and B. C. Subba Rao, *J. Org. Chem.*, **22**, 1136 (1957).

[92] G. Zweifel and H. C. Brown, Hydration of Olefins, Dienes and Acetylenes via Hydroboration, "Organic Reactions," John Wiley & Sons, Inc., New York, in press.

[93] H. C. Brown, C. Snyder, B. C. Subba Rao, and G. Zweifel, manuscript in preparation.

It is fortunate that the reaction appears to be almost specific for the boron-carbon linkage and is largely insensitive to other functional groupings. This was demonstrated by carrying out the above oxidation of tri-*n*-hexylborane in the presence of added reagents, such as 1-hexene, 3-hexyne, 1-hexyne, 1,3-cyclohexadiene, isobutyraldehyde, methylethylketone, *n*-butyl bromide, and acetonitrile. In no case was the quantitative production of 1-hexanol affected. Moreover, in all cases but isobutyraldehyde the added materials were demonstrated to be present in essentially unchanged form and concentration at the end of the oxidation. In the case of isobutyraldehyde, the recovery was only 83 per cent, evidently the result of some condensation under the alkaline conditions. However, by a more careful control of the alkalinity, it has proved possible to achieve excellent yields of aldehydes in the related oxidation of vinylboranes[74] (Chapter 16).

A detailed kinetic study of the reaction of alkaline hydrogen peroxide with trialkylboranes has yet to be made. However, there is available the related study by Kuivila and his co-workers of the reaction of benzene-boronic acid with hydrogen peroxide.[94-97] He proposes the mechanism

[94] H. G. Kuivila, *J. Am. Chem. Soc.*, **76**, 870 (1954).

[95] H. G. Kuivila, *J. Am. Chem. Soc.*, **77**, 4014 (1955).

[96] H. G. Kuivila and R. A. Wiles, *J. Am. Chem. Soc.*, **77**, 4830 (1955).

[97] H. G. Kuivila and A. G. Armour, *J. Am. Chem. Soc.*, **79**, 5659 (1957).

$$H_2O_2 + {}^-OH \rightleftharpoons HO_2^- + H_2O$$

$$\underset{\overset{|}{OH}}{\overset{\overset{C_6H_5}{|}}{HO-B}} + {}^-O_2H \longrightarrow \left[\underset{\overset{|}{OH}}{\overset{\overset{C_6H_5}{|}}{HO-B-OOH}} \right]^-$$

$$\left[\underset{\overset{|}{OH}}{\overset{\overset{C_6H_5}{|}}{HO-B-OOH}} \right]^- \longrightarrow \underset{\overset{|}{OH}}{HO-B-OC_6H_5} + OH^-$$

An identical mechanism, in three successive stages, is consistent with all the available data for the oxidation with alkaline hydrogen peroxide.[98] In this mechanism the organic group shifts with its pair of electrons from boron to oxygen. This is consistent both with the retention of configuration that is observed in the hydroboration-oxidation of cyclic olefins (Chapter 8) and the remarkable freedom from rearrangements that has been experienced to date.

Oxidation — Chromic Acid

The alkaline hydrogen peroxide oxidation of organoboranes provides an almost ideal means of converting the organometallic into the corresponding alcohol. These alcohols can be isolated and converted into other products, such as ketones and carboxylic acids, by standard methods. However, it was of interest to explore the possibility of a direct conversion of the organoborane into such oxidized products. Since aqueous chromic acid is the preferred reagent for the oxidation of secondary alcohols to ketones, we explored its action on organoboranes contained in the usual hydroboration solvents.[99]

One mention of a related reaction has appeared. Pappo has reported in a communication that the hydroboration of conessine, followed by chromic acid oxidation[100] in aqueous acetic acid,[101] yields 3-β-dimethyl-aminoconanin-6-one.

In our study, the olefin was hydroborated in diglyme, tetrahydrofuran, or ethyl ether. To the reaction mixture was added a small quantity of

[98] W. J. Wechter, *Chem. & Ind. (London)*, **1959**, 294.

[99] H. C. Brown and C. P. Garg, *J. Am. Chem. Soc.*, **83**, 2951 (1961).

[100] R. Pappo, *J. Am. Chem. Soc.*, **81**, 1010 (1959).

[101] R. Pappo, private communication, as reported in L. F. Fieser and M. Fieser, "Steroids," Reinhold Publishing Corporation, New York, 1959, p. 864.

water (to destroy residual hydride) followed by the slow addition of 10 per cent excess of aqueous chromic acid (from sodium dichromate and sulfuric acid) at 25 to 35°. After 2 hours the product was isolated. Yields of ketone in the range 65 to 85 per cent were realized.

Among the transformations realized were the following:

An unexpected byproduct of this research may be mentioned. The oxidation of secondary alcohols to ketones by aqueous chromic acid has long been a standard synthetic procedure. In seeking to improve this procedure, various workers have had recourse to solvents, such as acetic acid and acetone, solvents that are miscible with water and resist oxidation by chromic acid. However, in the course of the above study[99] we observed that the use of the immiscible solvent, ethyl ether, offered major advantages for chromic acid oxidations, especially in cases where the ketone is subject to epimerization.[102]

For example, the use of ethyl ether made possible the oxidation of *l*-menthol to *l*-menthone in 97 per cent yield, and isopinocampheol to isopinocamphone in 94 per cent yield (yields from gas chromatographic analysis). In each case only traces of the epimeric products were indicated, whereas the standard procedures resulted in the formation of several per cent of the epimers.[102]

Grignard-like Reactions — Metalation

The Grignard reagent is outstanding in the ease with which it transfers alkyl or aryl groups to the carbonyl or nitrile functions. However, it was early recognized that such transfers occur with difficulty, if at all, in the case of the organoboranes.[2,3]

Another major application of the Grignard reagent is in the synthesis of organometallics, such as diethylmercury and tetraethyllead. In these cases, the organoboranes offer promise of being highly useful.

For example, Honeycutt and Riddle observed that the treatment of a mixture of triethylborane and mercuric chloride in aqueous suspension with sodium hydroxide led to a rapid reaction at 70 to 80°, with the formation of a 95 per cent yield of diethylmercury. The reaction presumably involves the intermediate formation of mercuric oxide, and the reaction likewise proceeds when the oxide is utilized in place of the chloride.[103,104]

In the same way it has been possible to achieve the synthesis of tetraethyllead.

Thus far a detailed exploration of the scope of this reaction has not appeared. However, it has been shown that tri-*n*-hexylborane may be transformed into di-*n*-hexylmercury (55 per cent) and tetra-*n*-hexyllead. Similarly, triphenylborane has been converted into diphenylmercury (52 per cent).

[102] H. C. Brown and C. P. Garg, *J. Am. Chem. Soc.*, **83**, 2952 (1961).

[103] J. B. Honeycutt, Jr., and J. M. Riddle, *J. Am. Chem. Soc.*, **81**, 2593 (1959).

[104] J. B. Honeycutt, Jr., and J. M. Riddle, *J. Am. Chem. Soc.*, **82**, 3051 (1960).

It is quite remarkable that the transfer of alkyl groups proceeds so readily in aqueous systems. This opens up the possibility of a new general synthesis of organomercurials and related organometallics from olefins via the hydroboration reaction.

$$2RCH{=}CH_2 \xrightarrow{\text{HB}} 2RCH_2CH_2{-}B{\Big\langle}$$

$$NaOH \downarrow HgO$$

$$(RCH_2CH_2)_2Hg$$

Grignard-like Reactions — Cyclization

It was observed by Sommer and co-workers that γ-chloropropyltrichlorosilane undergoes rapid conversion to cyclopropane under the influence of alkali.[105]

$$Cl_3SiCH_2CH_2CH_2Cl + 4NaOH \longrightarrow \underset{H_2C\text{------}CH_2}{\overset{CH_2}{\triangle}} + 4NaCl + Si(OH)_4$$

The hydroboration of allyl chloride leads to the related derivatives, tri-(γ-chloropropyl)-borane and di-(γ-chloropropyl)-boron chloride.[106]

$$\underset{\overset{|}{Cl}}{CH_2CH}{=}CH_2 \xrightarrow{\text{HB}} (ClCH_2CH_2CH_2)_3B + (ClCH_2CH_2CH_2)_2BCl$$

Under the influence of aqueous alkali, these derivatives undergo an almost quantitative conversion to cyclopropane. Thus, 2.80 moles of cyclopropane were realized per mole of tri-(γ-chloropropyl)-borane and 1.90 moles of the cyclic hydrocarbon per mole of di-(γ-chloropropyl)-boron chloride.

The alkali is necessary to achieve a fast reaction at room temperature. For example, treatment of tri-(γ-chloropropyl)-borane at 100° with water alone produced only a 45 per cent conversion into cyclopropane in 1 week.

[105] L. H. Sommer, R. E. Van Strien, and F. C. Whitmore, *J. Am. Chem. Soc.,* **71,** 3056 (1949).

[106] M. F. Hawthorne and J. A. Dupont, *J. Am. Chem. Soc.,* **80,** 5830 (1958).

Presumably, the base coordinates with the organoborane, increasing the carbanion-like character of the boron-carbon linkage.

$$HO^- + \;>B-CH_2CH_2CH_2Cl \rightleftharpoons HO-\overset{\diagup}{\underset{}{B^-}}-CH_2$$

(with ring structure: $H_2C{-}{-}{-}CH_2$ and Cl)

$$HO-B< \;+ \quad \triangle \;(CH_2 / H_2C{-}{-}{-}CH_2) \;+ Cl^-$$

The use of phenyllithium in nonaqueous systems likewise facilitated the reaction, providing an 84 per cent yield.

This procedure has been explored as a general synthesis of cyclopropane derivatives.[107]   Methallyl chloride was hydroborated and the reaction product treated *in situ* with aqueous sodium hydroxide. A 71 per cent yield of methylcyclopropane was isolated. On a much smaller scale, phenylcyclopropane was obtained in a yield of 55 per cent and benzyl-cyclopropane in a yield of 45 per cent.

Since the cyclization reaction of the γ-chloropropylboranes appears to be essentially quantitative, the lower yields probably arise from one or more side reactions during the hydroboration stage. One such side reaction would be the attachment of the boron atom in significant amounts at the secondary or tertiary position, instead of at the primary.

$$\underset{\underset{Cl}{|}}{\overset{\overset{R}{|}}{CH_2C}}{=}CH_2 \xrightarrow{HB} \underset{\underset{B}{|}}{\overset{\overset{R}{|}}{ClCH_2CHCH_2}} + \underset{\underset{B}{|}}{\overset{\overset{R}{|}}{ClCH_2CCH_3}}$$

This orientation is unimportant in simple olefins, but can be markedly altered by aryl and halogen substituents (Chapter 7). It has been esti-mated that allyl chloride undergoes hydroboration to place 30 per cent of the boron at the secondary position.[108]

One means of controlling such undesired directive effects is the hydro-

[107] M. F. Hawthorne, *J. Am. Chem. Soc.*, **82**, 1886 (1960).
[108] P. Binger and R. Köster, *Tetrahedron Letters*, No. 4, 156 (1961).

boration with a dialkylborane, such as disiamylborane.[109]   Indeed the use of this reagent markedly improves the yield of cyclopropane realized in the combined hydroboration-cyclization of allyl chloride.[110]

Binger and Köster have proposed the combined use of dialkylboranes for the hydroboration stage and sodium hydride for the cyclization stage. This permits regeneration of the dialkylborane for subsequent utilization in a cyclic process.[108]

$$R_2BH + H_2C=CHCH_2Cl \longrightarrow R_2BCH_2CH_2CH_2Cl$$

$$NaH + R_2BCH_2CH_2CH_2Cl \longrightarrow Na^+ \left[\overset{H}{R_2\ddot{B}CH_2CH_2CH_2Cl} \right]^-$$

$$Na^+ \left[\overset{H}{R_2\ddot{B}CH_2CH_2CH_2Cl} \right]^- \longrightarrow R_2BH + (CH_2)_3 + NaCl$$

Attempts to bring about a similar synthesis of cyclobutane from 4-chloro-1-butene were unsuccessful.[107,108]   The hydroboration proceeded normally, but the product resisted cyclization.

Coupling

The reaction of Grignard reagents with silver bromide provides a convenient means of bringing about the formation of carbon-carbon bonds, with yields of coupled products in the range 40 to 60 per cent.[111]   Johnson and his co-workers reported that the treatment of n-butylboronic acid and n-hexylboronic acid with ammoniacal silver oxide brings about a similar coupling of the alkyl groups to form n-octane and n-dodecane, respectively.[8,9]   It therefore appeared of interest to explore the possibility of achieving the coupling reaction in the trialkylboranes.

Neither silver bromide nor Tollens reagent yielded any significant quantity of product.   Silver oxide was somewhat more favorable, bringing about the formation from triethylborane of 9 per cent n-butane, 5 per cent ethylene, and 15 per cent ethane in 18 hours at room temperature.

The addition of sodium hydroxide exerted a remarkable effect on the reaction, bringing about complete conversion to products (72 per cent n-butane, 9 per cent ethylene, and 9 per cent ethane) in a matter of minutes at either 25 or 0°.[112]   Gold compounds served similarly.[112]

[109] H. C. Brown and G. Zweifel, *J. Am. Chem. Soc.*, **82**, 3222 (1960).

[110] H. C. Brown and K. Keblys, unpublished observations.

[111] J. H. Gardner and P. Borgstrom, *J. Am. Chem. Soc.*, **51**, 3375 (1929).

[112] H. C. Brown, N. C. Hébert, and C. H. Snyder, *J. Am. Chem. Soc.*, **83**, 1001 (1961).

It proved possible to achieve the coupling reaction directly in the hydroboration flask, permitting a combined hydroboration-coupling procedure as a convenient route to the products.[113] In a typical procedure 100 mmoles of 1-hexene is hydroborated in diglyme with sodium borohydride and boron trifluoride. A small quantity of water is added to destroy residual hydride, followed by 120 ml of 2.0 M aqueous potassium hydroxide. The reaction mixture is cooled to 0°, and 24.0 ml of a 5.0 M solution of silver nitrate is added over 10 minutes. After 1 hour at 0°, the product is isolated. A 70 per cent yield of the coupled product is obtained.

The coupling reaction for more hindered olefins, such as 2-methyl-1-pentene, gives better yields in methanol solution, presumably because of the greater difficulty of reaction of potassium hydroxide with the more hindered organoborane in the two-phase aqueous system.

The reaction appears to be widely applicable. Terminal olefins undergo coupling with yields in the range 60 to 80 per cent. Internal olefins form the coupled products in yields of 35 to 50 per cent.

$$
\begin{array}{c}
\quad\ \ C \qquad\qquad\qquad\qquad C \qquad\qquad\qquad\quad C \\
\quad\ \ | \qquad\qquad\qquad\qquad\ | \qquad\qquad\qquad\quad\ | \\
C-C-C\!=\!C \longrightarrow C-C-C-C-C-C-C-C \\
\quad\ \ | \qquad\qquad\qquad\qquad\ | \qquad\qquad\qquad\quad\ | \\
\quad\ \ C \qquad\qquad\qquad\qquad C \qquad\qquad\qquad\quad C
\end{array}
$$

$$
\begin{array}{c}
\qquad\qquad C \qquad\qquad\qquad\qquad\quad C \qquad\quad C \\
\qquad\qquad | \qquad\qquad\qquad\qquad\quad | \qquad\quad\ | \\
C-C-C-C\!=\!C \longrightarrow C-C-C-C-C-C-C-C-C-C
\end{array}
$$

$$
\begin{array}{c}
\ \ C \quad\ C \qquad\qquad\qquad\quad C \quad\ C \qquad\qquad C \quad\ C \\
\ \ | \quad\ | \qquad\qquad\qquad\quad | \quad\ | \qquad\qquad | \quad\ | \\
C-C-C-C\!=\!C \longrightarrow C-C-C-C-C-C-C-C-C-C \\
\ \ | \qquad\qquad\qquad\qquad\quad\ | \qquad\qquad\qquad\qquad\ | \\
\ \ C \qquad\qquad\qquad\qquad\quad\ C \qquad\qquad\qquad\qquad\ C
\end{array}
$$

$$
\begin{array}{c}
\qquad\qquad\qquad\qquad C \quad C \\
\qquad\qquad\qquad\qquad | \quad | \\
C-C\!=\!C-C \longrightarrow C-C-C-C-C-C
\end{array}
$$

$$
\begin{array}{c}
\ \ C \qquad\qquad\qquad\qquad C \quad C \quad C \quad C \\
\ \ | \qquad\qquad\qquad\qquad | \quad | \quad | \quad | \\
C-C\!=\!C-C \longrightarrow C-C-C-C-C-C
\end{array}
$$

[113] H. C. Brown and C. H. Snyder, *J. Am. Chem. Soc.*, **83**, 1001 (1961).

The reaction has also been explored as a means of joining two different alkyl groups.[114]   In a reaction involving a statistical coupling of two different groups, R and R′, the maximum yield of the desired product can only be 50 per cent of the coupled material, R_2, R—R′, R_2'.   However, there is the possibility of improving the yield over this limit by utilizing a large excess of a relatively cheap olefin to achieve a more complete conversion of a second, more expensive olefin into the desired product.   The practicality of this approach was demonstrated.[114]

The power of this new synthetic procedure is indicated by the following syntheses:

[114] H. C. Brown, C. Verbrugge, and C. H. Snyder, *J. Am. Chem. Soc.*, **26**, 1001 (1961).

$$C — C — C — C — C = C$$
$$\qquad\qquad\quad |$$
$$\qquad\qquad\quad C$$
$$+ \quad C — C = C — C \qquad \longrightarrow \quad \text{2,3-dimethylnonane}$$

The hydroboration reaction can tolerate many different functional groups (Chapter 19) not compatible with the Grignard reagent. Consequently this simple formation of carbon-carbon bonds should have very wide applicability in synthetic chemistry.

It is probable that the reaction proceeds through the formation of the silver alkyl. These are unstable under the reaction conditions, presumably breaking down into silver and the free radical.[115] The free radicals present in high concentration unite to form the product.

It is apparent that this reaction offers promise as a new, simple route to free radical reactions. For example, the presence of carbon tetrachloride in the reaction mixture diverts the reaction from dimerization to the formation of the corresponding alkyl chloride.[112,116] The scope of this new entry into free-radical chemistry is being explored.

Conclusion

As was pointed out earlier, the systematic study of the chemistry of the organoboranes is of very recent origin. Nevertheless, a number of highly interesting reactions of considerable value for synthetic chemistry have been uncovered. We may be confident that the continued study of this essentially virgin area will continue to provide new chemical developments of major interest and utility. At that stage the organoboranes will take their place with the organomagnesium compounds and the complex hydrides as another major tool of the synthetic chemist.

[115] C. E. H. Bawn and R. Johnson, *J. Chem. Soc.*, **1960**, 3923.
[116] H. C. Brown and D. Burton, unpublished observations.

4 | Borohydride Chemistry

In 1942 there were available two synthetic routes to diborane. The first, developed by Stock in 1912, involved the preparation and hydrolysis of magnesium boride followed by cracking the higher boron hydrides thus obtained.[1] The process was exceedingly tedious — several months effort was required to obtain a few grams of product.

The second, introduced by Schlesinger and Burg in 1931, was a vast improvement. It utilized the interaction of hydrogen and boron trichloride in the silent electric discharge to produce a partially hydrogenated boron chloride intermediate which could be disproportionated into boron trichloride and diborane.[2] This procedure made possible the preparation of diborane at the rate of several tenths of a gram per day.

As long as diborane was only a laboratory curiosity, these quantities were adequate, and there was little incentive to investigate improved procedures. For example, my doctorate thesis involved a detailed study of the reaction of diborane with a number of aldehydes, ketones, esters, and acid chlorides.[3] It required a total of 500 ml of diborane gas, or approximately one-half of a gram.

Early Studies

This situation was altered by war research. We were requested to undertake the synthesis of new volatile compounds of uranium. This led to the preparation of uranium(IV) borohydride, $U(BH_4)_4$.[4] It possessed interesting

[1] A. Stock and C. Massenez, *Ber.*, **45**, 3539 (1912).

[2] H. I. Schlesinger and A. B. Burg, *J. Am. Chem. Soc.*, **53**, 4321 (1931).

[3] H. C. Brown, H. I. Schlesinger, and A. B. Burg, *J. Am. Chem. Soc.*, **61**, 673 (1939).

[4] H. I. Schlesinger and H. C. Brown, *J. Am. Chem. Soc.*, **75**, 219 (1953).

properties, and relatively large quantities were required for further investigation. Obviously the older procedures proceeding through diborane, obtained by the arc process, to the metal borohydrides[5–7] were inadequate. Consequently, a research program was undertaken to develop more convenient, more practical routes to these substances.

We soon discovered that boron trifluoride etherate in ether solution reacts readily with finely divided lithium hydride in suspension to give diborane.[8] Moreover, in the same solvent, diborane and lithium hydride react to form lithium borohydride, providing a new simple route to this valuable intermediate. The lithium borohydride is soluble in ether and reacts rapidly with boron halides in that solvent to liberate diborane quantitatively.[9]

$$6LiH + 8BF_3 \xrightarrow{EE} B_2H_6 + 6LiBF_4$$

$$2LiH + B_2H_6 \xrightarrow{EE} 2LiBH_4$$

$$3LiBH_4 + 4BF_3 \xrightarrow{EE} 2B_2H_6 + 3LiBF_4$$

$$3LiBH_4 + BCl_3 \xrightarrow{EE} 2B_2H_6 + 3LiCl$$

Unfortunately, the relative scarcity of lithium metal forced us to turn our attention away from the convenient lithium derivatives to sodium. Sodium hydride proved to be much more sluggish in its reactions. It could be made to react with boron trifluoride etherate only at elevated temperatures, preferably in a ball mill or similar device which kept the surface of the insoluble hydride exposed and active. (At that time, 1942–1943, the solvents that can now be used to facilitate these reactions were not available.)

Fortunately, it proved possible to circumvent these difficulties by utilizing an addition compound of sodium hydride with methyl borate,

[5] H. I. Schlesinger, R. T. Sanderson, and A. B. Burg, *J. Am. Chem. Soc.*, **62**, 3421 (1940).

[6] A. B. Burg and H. I. Schlesinger, *J. Am. Chem. Soc.*, **62**, 3425 (1940).

[7] H. I. Schlesinger and H. C. Brown, *J. Am. Chem. Soc.*, **62**, 3429 (1940).

[8] H. I. Schlesinger, H. C. Brown, J. R. Gilbreath, and J. J. Katz, *J. Am. Chem. Soc.*, **75**, 195 (1953).

[9] H. I. Schlesinger, H. C. Brown, H. R. Hoekstra, and L. R. Rapp, *J. Am. Chem. Soc.*, **75**, 199 (1953).

sodium trimethoxyborohydride.[10]   This interesting substance is readily prepared by heating sodium hydride with excess methyl borate under reflux.   The reaction proceeds with a remarkable expansion in the volume of the solid in the reaction vessel and is readily followed by observing the increase in volume.   The product behaves as a form of active or soluble sodium hydride.   Reaction readily occurs with diborane to form sodium borohydride and with boron halides to form diborane.

$$NaH + B(OCH_3)_3 \longrightarrow NaBH(OCH_3)_3$$

$$2NaBH(OCH_3)_3 + B_2H_6 \longrightarrow 2NaBH_4 + 2B(OCH_3)_3$$

$$6NaBH(OCH_3)_3 + 8BF_3 \longrightarrow B_2H_6 + 6NaBF_4 + 6B(OCH_3)_3$$

$$6NaBH(OCH_3)_3 + 2BCl_3 \longrightarrow B_2H_6 + 6NaCl + 6B(OCH_3)_3$$

At elevated temperatures methyl borate reacts with sodium hydride to form sodium borohydride and sodium methoxide.[11]   Since these are readily separated by the use of ammonia or isopropylamine as a selective solvent for the borohydride, this reaction provides a convenient route to sodium borohydride.

$$4NaH + B(OCH_3)_3 \xrightarrow{\text{250}°} NaBH_4 + 3NaOCH_3$$

From sodium borohydride it proved possible to proceed to other metal borohydrides.[12]

$$LiCl + NaBH_4 \xrightarrow{\text{isopropylamine}} LiBH_4 + NaCl \downarrow$$

$$BeBr_2 + 2NaBH_4 \xrightarrow{\Delta} Be(BH_4)_2 \uparrow + 2NaBr$$

$$AlCl_3 + 3NaBH_4 \longrightarrow Al(BH_4)_3 \uparrow + 3NaCl$$

[10] H. C. Brown, H. I. Schlesinger, I. Sheft, and D. M. Ritter, *J. Am. Chem. Soc.*, **75**, 192 (1953).

[11] H. I. Schlesinger, H. C. Brown, and A. E. Finholt, *J. Am. Chem. Soc.*, **75**, 205 (1953).

[12] H. I. Schlesinger, H. C. Brown, and E. K. Hyde, *J. Am. Chem. Soc.*, **75**, 209 (1953).

The alkali metal hydride route was later extended successfully to the synthesis of lithium aluminum hydride,[13] as well as the previously discovered[14] aluminum hydride, and made possible a rapid exploration of the synthetic chemistry of metal hydrides and double hydrides.[15,16] The developments in this area have had far-reaching consequences, both in inorganic and in organic chemistry,[17,18] and it is not yet possible to visualize their full consequences.

Recent Studies

These early studies were carried out under emergency conditions, with immediate practical objectives. As a consequence, the research could not be planned and executed with the care and thoroughness of normal academic research. Accordingly, it appeared desirable to initiate a new research program in this area involving a systematic study of the reaction of Lewis acids of boron with the alkali metal hydrides and borohydrides.

$$MH + BX_3 \longrightarrow M^+BHX_3^-$$

$$MH + BH_3 \longrightarrow M^+BH_4^-$$

$$MH + BR_3 \longrightarrow M^+BHR_3^-$$

$$MH + B(OR)_3 \longrightarrow M^+BH(OR)_3^-$$

This research program was originally initiated with the primary objective of making the older, relatively fragmentary studies more complete, eliminating annoying gaps in our knowledge. However, it has turned out to be far more fruitful than anticipated. Although the program is far from complete, it has not only eliminated many of these gaps, but it has also turned up new phenomena of both theoretical and practical interest.

[13] A. E. Finholt, A. C. Bond, Jr., K. E. Wilzbach, and H. I. Schlesinger, *J. Am. Chem. Soc.*, **69**, 2692 (1947).

[14] O. Stecher and E. Wiberg, *Ber.*, **75**, 2003 (1942).

[15] E. Wiberg, *Angew. Chem.*, **65**, 16 (1953).

[16] G. D. Barbaras, C. Dillard, A. E. Finholt, T. Wartik, K. E. Wilzbach, and H. I. Schlesinger, *J. Am. Chem. Soc.*, **73**, 4585 (1951).

[17] K. Ziegler, *Angew. Chem.*, **68**, 721 (1956).

[18] K. Ziegler, Organo-Aluminum Compounds, in H. Zeiss (ed.), "Organometallic Chemistry," Chap. 5, American Chemical Society Monograph Series, Reinhold Publishing Corporation, New York, 1960.

The early attempts to react metal hydrides and boron halides in the absence of solvents encountered difficulties.[8,19] The use of ethyl ether eliminated these difficulties for the reactions involving lithium derivatives but not for sodium. Presumably, this difference in the behavior of the two systems arises from the solubility of many lithium salts in ethyl ether, whereas the sodium compounds exhibit little or no solubility in this solvent. Consequently, a search was undertaken for inert liquids that would serve as solvents for such intermediates as sodium borohydride, trimethoxy-borohydride, and tetramethoxyborohydride.[20] Of the materials examined, tetrahydrofuran and the dimethylether of diethyleneglycol (diglyme) appeared to be the most promising and were selected for detailed study. The results are summarized in Table 4-1.

Table 4-1

Solubility of Sodium Borohydride and Related Products in Ether Solvents

	Ethyl ether	Tetrahydrofuran	Diglyme
$NaBH_4$	Insol.	Slightly sol.	Sol.
$NaBH(OCH_3)_3$	Insol.	Very sol.	Sol.
$NaB(OCH_3)_4$	Insol.	Very sol.	Insol.
$NaBF_4$	Insol.	Insol.	Sol.
B_2H_6	Slightly sol.	Very sol.	Slightly sol.

The solubility of sodium borohydride in diglyme exhibits unusual characteristics. At 100° the solubility is quite low, approximately 0.1 M. The solubility increases as the temperature is lowered, reaching a maximum solubility of approximately 3 M at 40°. The solubility then decreases to approximately 0.3 M at 0°. At the lower temperatures a 1:1 solvate of diglyme and sodium borohydride is in equilibrium with the solution. At the higher temperatures, the unsolvated salt separates from the solution.

The dimethylether of triethyleneglycol (triglyme) is also a useful solvent for sodium borohydride. In this solvent the solubility is high, in the neighborhood of 3 M, and does not vary significantly over the temperature range 0 to 100°. Consequently, it is useful when a higher-boiling solvent is desired (diglyme, b.p. 162°; triglyme, b.p. 216°), or when one wishes to realize a high concentration of sodium borohydride at either relatively high or low temperatures.

[19] J. Goubeau and R. Bergmann, *Z. anorg. u. allgem. Chem.*, **263**, 69 (1950).
[20] H. C. Brown, E. J. Mead, and B. C. Subba Rao, *J. Am. Chem. Soc.*, **77**, 6209 (1955).

In diglyme, sodium hydride readily reacts with boron trifluoride to give a quantitative yield of diborane. Likewise, in this solvent, sodium hydride reacts with diborane to form sodium borohydride, and the latter reacts with boron trifluoride to liberate diborane.[21] Consequently, all the reactions previously exhibited by lithium hydride[8] and borohydride[9] in ether solvents are readily duplicated by the corresponding sodium derivatives in the solvent diglyme.

$$6NaH + 8BF_3 \xrightarrow[25°]{DG} B_2H_6 + 6NaBF_4$$

$$2NaH + B_2H_6 \xrightarrow[25°]{DG} 2NaBH_4$$

$$3NaBH_4 + 4BF_3 \xrightarrow[25°]{DG} 2B_2H_6 + 3NaBF_4$$

Under these conditions boron trichloride does not react with sodium hydride at any convenient rate. Indeed, it attacks the solvent to form methyl chloride. However, boron trichloride does react smoothly and quantitatively with sodium borohydride in diglyme solution. Consequently, it is possible to effect a two-stage reaction to bring about the preparation of diborane from sodium hydride and boron trichloride.

$$6NaH + 3B_2H_6 \longrightarrow 6NaBH_4$$

$$6NaBH_4 + 2BCl_3 \longrightarrow 4B_2H_6 + 6NaCl$$

$$\overline{6NaH + 2BCl_3 \longrightarrow B_2H_6 + 6NaCl}$$

The addition of lithium chloride or bromide to a solution of sodium hydride in diglyme produces a precipitate of the sodium halide and a solution of lithium borohydride. Similarly, the addition of magnesium halide results in the formation of magnesium borohydride in solution.[20] Lithium and magnesium borohydrides are much more powerful reducing agents than sodium borohydride,[22,23] so that this addition of the appro-

[21] H. C. Brown and P. A. Tierney, *J. Am. Chem. Soc.*, **80**, 1552 (1958).

[22] R. F. Nystrom, S. W. Chaikin, and W. G. Brown, *J. Am. Chem. Soc.*, **71**, 3245 (1949).

[23] J. Kollonitsch, O. Fuchs, and V. Gabor, *Nature*, **173**, 125 (1954).

priate salt provides a simple means of enhancing the reducing power of the borohydride solution.

$$LiCl + NaBH_4 \xrightarrow{\text{DG}} LiBH_4 + NaCl \downarrow$$

$$MgCl_2 + 2NaBH_4 \xrightarrow{\text{DG}} Mg(BH_4)_2 + 2NaCl \downarrow$$

The addition of anhydrous aluminum chloride to sodium borohydride in diglyme solution in the stoichiometric ratio ($3NaBH_4 + AlCl_3$) gives a clear solution. The absence of a precipitate of sodium chloride indicates that the reaction has not proceeded to the formation of· aluminum borohydride.

$$AlCl_3 + 3NaBH_4 \xrightarrow{\text{DG}} Al(BH_4)_3 + 3NaCl \downarrow$$

Yet the solution exhibits exceedingly powerful reducing action, approaching that of lithium aluminum hydride itself.[24] Possibly the reaction does involve the formation of a small equilibrium concentration of aluminum borohydride or related derivative, so that the solubility of sodium chloride is not exceeded. On this basis, sufficient sodium chloride to precipitate is formed only as the equilibrium is shifted by reaction with the organic component undergoing reduction.

$$AlCl_3 + 3NaBH_4 \xleftrightarrow{\text{DG}} Al(BH_4)_3 + 3NaCl$$

Thus far the results were no more than we had expected to achieve with a stable solvent for sodium borohydride. However, we encountered a number of phenomena that were quite unexpected.

Treatment of sodium hydride in diglyme with excess diborane resulted in the absorption of a full mole of diborane. Likewise, sodium borohydride in diglyme absorbed an additional half-molar equivalent of diborane. Evidently, there must exist in diglyme solution a species, $NaBH_4 \cdot BH_3$ or NaB_2H_7, sodium diborohydride.

$$NaH + B_2H_6 \longrightarrow NaB_2H_7$$

$$NaBH_4 + \tfrac{1}{2}B_2H_6 \xrightleftharpoons{} NaB_2H_7$$

The reaction is reversible. At elevated temperatures the additional diborane can be recovered. Consequently, the formation of sodium

[24] H. C. Brown and B. C. Subba Rao, *J. Am. Chem. Soc.*, **78**, 2582 (1956).

diborohydride provides a convenient means of absorbing, storing, and purifying diborane in laboratory research.

Because of the formation of this species, the addition of boron trifluoride etherate to a diglyme solution of sodium borohydride does not liberate diborane until nearly half the boron trifluoride has been added. Then diborane is evolved relatively rapidly with the continued addition of the fluoride. Consequently, in order to generate diborane smoothly and continuously, it is preferable to add the solution of sodium borohydride to the boron trifluoride etherate in diglyme. In that event, the diborane evolved corresponds precisely to the amount of sodium borohydride that has been introduced into the generator flask.

One of the preferred hydroboration procedures involves the slow addition of boron trifluoride etherate to a mixture of sodium borohydride and the olefin in diglyme.[25] In this procedure the rapid and reversible formation of sodium diborohydride provides an internal safety device to minimize any loss of diborane from the solution.

In diglyme solution lithium borohydride likewise reacts with additional diborane. The diborohydride ion appears to be representative of a large class of singly-bridged compounds formed by the boron halides and boron hydrides,[26] related to the more commonly recognized doubly-bridged molecules, such as diborane and aluminum bromide.

Another unexpected phenomenon was noted in the reaction of boron trichloride with sodium borohydride.[21] A quantitative yield of diborane is obtained with the theoretical quantity of boron trichloride. However, the use of excess boron trichloride results in a decreased yield, and with two molar equivalents of boron trichloride no diborane is evolved. These phenomena arise from the ready formation of a new class of compounds, the chloroborane etherates.[27]

$$3NaBH_4 + BCl_3 : OR_2 \longrightarrow 2B_2H_6 + 3NaCl + R_2O$$

$$3NaBH_4 + 3BCl_3 : OR_2 + 3OR_2 \longrightarrow 6ClH_2B : OR_2 + 3NaCl$$

$$3NaBH_4 + 9BCl_3 : OR_2 + 3OR_2 \longrightarrow 12Cl_2HB : OR_2 + 3NaCl$$

$$B_2H_6 + BCl_3 : OR_2 \longrightarrow \text{no reaction}$$

$$B_2H_6 + BCl_3 : OR_2 + 2OR_2 \longrightarrow 3ClH_2B : OR_2$$

$$B_2H_6 + 4BCl_3 : OR_2 + 2OR_2 \longrightarrow 6Cl_2HB : OR_2$$

[25] H. C. Brown and B. C. Subba Rao, *J. Am. Chem. Soc.*, **81**, 6428 (1959).

[26] H. C. Brown, P. F. Stehle, and P. A. Tierney, *J. Am. Chem. Soc.*, **79**, 2020 (1957).

[27] H. C. Brown and P. A. Tierney, *J. Inorg. & Nuclear Chem.*, **9**, 51 (1959).

Although sodium borohydride is only slightly soluble in tetrahydrofuran, the solubility is sufficient to achieve certain reactions. For example, the careful treatment of a suspension of sodium borohydride in this solvent at 0° with hydrogen chloride provides a solution of diborane in tetrahydrofuran.[28] Similarly, boron trifluoride etherate converts a suspension of sodium borohydride into a solution of diborane in the solvent.[29]

$$NaBH_4 + HCl \xrightarrow[0°]{THF} BH_3:THF + H_2 + NaCl \downarrow$$

$$3NaBH_4 + 4BF_3 \xrightarrow[0°]{THF} 4BH_3:THF + 3NaBF_4 \downarrow$$

Attention should also be called to the convenient conversion of the insoluble potassium borohydride into the soluble and reactive lithium borohydride by stirring a suspension of lithium chloride and potassium borohydride in tetrahydrofuran.[30]

$$KBH_4(s) + LiCl(s) \xrightarrow{THF} LiBH_4(soln) + KCl(s)$$

Even ethyl ether may be utilized as a solvent. Sodium borohydride is insoluble in this solvent and reactions involving this borohydride are sluggish. However, zinc chloride reacts with sodium borohydride to provide a solution of zinc borohydride in ethyl ether. This solution can be utilized for the generation of diborane or other reactions requiring soluble borohydride. In some cases it has proved satisfactory to utilize catalytic quantities of zinc chloride.[31,32]

$$2NaBH_4(s) + ZnCl_2(soln) \longrightarrow Zn(BH_4)_2(soln) + 2NaCl$$

[28] H. C. Brown and G. Zweifel, *J. Am. Chem. Soc.*, **81**, 4106 (1959).

[29] H. C. Brown and L. J. Murray, unpublished observations.

[30] R. Paul and N. Joseph, *Bull. soc. chim. France*, **1952**, 550.

[31] H. C. Brown, K. J. Murray, L. J. Murray, J. A. Snover, and G. Zweifel, *J. Am. Chem. Soc.*, **82**, 4233 (1960).

[32] N. Joseph, XVIII International Congress of Pure and Applied Chemistry, Munich, Sept. 1959.

Finally, it should be pointed out that lithium aluminum hydride is readily soluble in ethyl ether, and such solutions may readily be utilized for the preparation of diborane.[13,33]

It was early shown that in ether solution trimethylborane readily reacts with lithium hydride to form lithium trimethylborohydride.[10] Triethylborane proved capable of reacting with sodium hydride in the absence of any solvent.[10] A detailed study of these reactions has demonstrated that they are general, occurring with a wide range of organoboranes.[34]

A number of points of interest should be mentioned. Lithium hydride dissolves in ether solvents (ethyl ether, n-butyl ether, tetrahydrofuran, and diglyme) containing trialkylborane. At elevated temperatures the addition compounds in solution dissociate, giving off the trialkylborane and precipitating lithium hydride. Consequently, a mixture of trialkylborane and an ether may be considered a solvent for lithium hydride from which lithium hydride may be recrystallized.

$$LiH(s) + BR_3(soln) \underset{}{\overset{ether}{\rightleftharpoons}} LiBHR_3(soln)$$

In diglyme or tetrahydrofuran solution, 2 moles of trimethylborane are absorbed by 1 mole of lithium hydride or sodium hydride. Consequently, the solutions presumably contain diborohydride derivatives related to the simple diborohydrides mentioned earlier.

$$LiH + 2(CH_3)_3B \overset{DG}{\rightleftharpoons} Li^+[(CH_3)_3BHB(CH_3)_3]^-$$

$$NaH + 2(CH_3)_3B \overset{THF}{\rightleftharpoons} Na^+[(CH_3)_3BHB(CH_3)_3]^-$$

The reaction product of sodium hydride and triethylborane is an oil, easily soluble in a number of relatively nonpolar solvents. Since the substance is presumably $Na^+BH(C_2H_5)_3^-$, the existence of a sodium salt, liquid at room temperature, is of considerable interest. Finally, these borohydrides are considerably more powerful reducing agents than those of the parent unsubstituted borohydrides, and they permit a number of interesting selective reductions.

[33] I. Shapiro, H. G. Weiss, M. Schmich, S. Skolnik, and G. B. L. Smith, J. Am. Chem. Soc., 74, 901 (1952).

[34] H. C. Brown and A. Khuri, unpublished observations.

In the absence of solvents, lithium and sodium trimethoxyborohydride are obtained by the reaction of methyl borate with the corresponding alkali metal hydrides.[10]   Since sodium trimethoxyborohydride is soluble in tetrahydrofuran, we attempted to facilitate the reaction by carrying it out in that solvent.   Unexpectedly, the product did not consist of the simple addition compound, but of a precipitate of sodium borohydride and a solution of sodium tetramethoxyborohydride.[35]

$$NaH + B(OCH_3)_3 \longrightarrow NaBH(OCH_3)_3$$
$$4NaBH(OCH_3)_3 \xrightarrow{\text{THF}} NaBH_4 \downarrow + 3NaB(OCH_3)_4 (soln)$$

In diglyme, sodium borohydride is soluble and sodium tetramethoxyborohydride is insoluble.   In this solvent, the disproportionation leads to a precipitate of the latter and a solution of the former.[34]

$$4NaBH(OCH_3)_3 \xrightarrow{\text{DG}} NaBH_4 (soln) + 3NaB(OCH_3)_4 \downarrow$$

It proved possible to synthesize sodium triisopropoxyborohydride and tri-t-butoxyborohydride.[36]   In these cases disproportionation is either slow or nonexistent.   Consequently, it proved possible to examine the reducing action of these derivatives in solution.   They proved to be considerably more powerful than the parent borohydrides.

Amine-Boranes

The high reactivity of diborane is presumably the result of its ready transformation into the reactive borane moiety with its deficiency of electrons.   Tertiary amines, such as trimethylamine and pyridine, react with diborane to form the addition compounds, trimethylamine-borane[37] and pyridine-borane,[38] which are coordinatively satisfied.

[35] H. C. Brown, E. J. Mead, and P. A. Tierney, *J. Am. Chem. Soc.*, **79**, 5400 (1957).

[36] H. C. Brown, E. J. Mead, and C. J. Shoaf, *J. Am. Chem. Soc.*, **78**, 3613 (1956).

[37] H. I. Schlesinger, and A. B. Burg, *J. Am. Chem. Soc.*, **60**, 290 (1938).

[38] H. C. Brown, H. I. Schlesinger, and S. Z. Cardon, *J. Am. Chem. Soc.*, **64**, 325 (1942).

$$(CH_3)_3N + \tfrac{1}{2}(BH_3)_2 \longrightarrow (CH_3)_3N:BH_3$$

The products are relatively stable substances which can be handled in the presence of air and moisture.

Recently, a new synthetic procedure has been described for the synthesis of triethylamine-borane. This procedure involves the preparation of triethylborane from triethylaluminum, followed by the hydrogenation of triethylamine-triethylborane.[39]

$$(C_2H_5)_3Al + B(OCH_3)_3 \longrightarrow (C_2H_5)_3B + Al(OC_2H_5)_3$$

$$(C_2H_5)_3B + (C_2H_5)_3N \longrightarrow (C_2H_5)_3N:B(C_2H_5)_3$$

$$(C_2H_5)_3N:B(C_2H_5)_3 + 3H_2 \longrightarrow (C_2H_5)_3N:BH_3 + 3C_2H_6$$

It has been established that triethylamine-borane can be transformed into sodium borohydride by procedures analogous to those previously developed for diborane itself.

$$NaH + (C_2H_5)_3N:BH_3 \longrightarrow NaBH_4 + (C_2H_5)_3N$$

Conclusion

This background in the synthetic chemistry of diborane, the metal borohydrides and the amine-boranes should facilitate following the many procedures that have been developed for achieving the hydroboration of olefins, discussed in the following chapter.

[39] R. Köster, *Angew. Chem.*, **69**, 94 (1957).

5 | Hydroboration Procedures

Sodium borohydride is the most readily available of the complex borohydrides. Consequently, our early studies of the hydroboration reaction were largely based on this material. It is essentially insoluble in ethyl ether, only slightly soluble in tetrahydrofuran, but is readily soluble in the solvent diglyme (diethyleneglycol dimethyl ether).[1] Diglyme (DG) was therefore the solvent of choice in bringing about the aluminum chloride–catalyzed reaction of sodium borohydride and olefins.[2,3]

$$9RCH{=}CH_2 + 3NaBH_4 + AlCl_3 \xrightarrow{DG} 3(RCH_2CH_2)_3B + 3NaCl + AlH_3$$

This reaction suffers from the disadvantage that only three of the four "hydrides" of sodium borohydride are utilized for hydroboration. The fourth is lost in the form of aluminum hydride. The use of boron trifluoride etherate or boron trichloride circumvents this difficulty.[4,5]

$$12RCH{=}CH_2 + 3NaBH_4 + 4BF_3{:}O(C_2H_5)_2 \xrightarrow{DG}$$
$$4(RCH_2CH_2)_3B + 3NaBF_4 + 4(C_2H_5)_2O$$

$$12RCH{=}CH_2 + 3NaBH_4 + BCl_3 \xrightarrow{DG} 4(RCH_2CH_2)_3B + 3NaCl$$

---

[1] H. C. Brown, E. J. Mead, and B. C. Subba Rao, *J. Am. Chem. Soc.*, **77**, 6209 (1955).
[2] H. C. Brown and B. C. Subba Rao, *J. Am. Chem. Soc.*, **78**, 5694 (1956).
[3] H. C. Brown and B. C. Subba Rao, *J. Am. Chem. Soc.*, **81**, 6423 (1959).
[4] H. C. Brown and B. C. Subba Rao, *J. Org. Chem.*, **22**, 1136 (1957).
[5] H. C. Brown and B. C. Subba Rao, *J. Am. Chem. Soc.*, **81**, 6428 (1959).

Boron trichloride possesses the advantage that only 1 mole is required to achieve the hydroboration of 12 moles of olefin, whereas 4 moles of boron trifluoride etherate are required for the same transformation. On the other hand, boron trichloride is a gas (b.p. 12°) difficult to handle quantitatively by the usual laboratory procedures. Moreover, it is a highly reactive substance which rapidly splits many ethers.[6,7] Boron trifluoride etherate is a relatively stable substance toward ethers. It is a liquid (b.p. 125°), easily purified and measured. Consequently, the addition of boron trifluoride etherate to a mixture of the olefin and sodium borohydride in diglyme became the procedure of choice for the hydroboration of olefins.

Occasionally, diglyme (b.p. 162°) may distill too close to the reaction product to permit its convenient purification. In that event, it is a simple matter to shift to triglyme (triethyleneglycol dimethyl ether, b.p. 216°). Actually, these solvents are much more soluble in water than in ethyl ether, so that the reaction mixture can be poured into water and the product selectively extracted into ethyl ether.

In some cases the presence of ethyl ether, from the boron trifluoride etherate, has proved to be undesirable. Thus it frequently interferes with the gas-chromatographic analysis of olefins. In such cases the boron trifluoride etherate is simply dissolved in the diglyme and the diethyl ether is removed under vacuum. The solution of boron trifluoride diglymate is then added to the reaction mixture in the usual manner to achieve hydroboration.[8]

Finally, it appeared desirable to have a procedure available which would permit the conversion of olefins into the organoboranes without the concurrent formation of a salt. The rapid reaction of diborane with olefins[4,5] made this possible. Diborane is generated by adding a solution of sodium borohydride in diglyme to a solution of boron trifluoride etherate in diglyme.[7] The gas is immediately conducted into a solution of the olefin in an appropriate ether, such as diglyme, tetrahydrofuran, or ethyl ether. The high solubility of diborane in tetrahydrofuran makes it the solvent of choice for this procedure (external hydroboration).

$$6RCH\!\!=\!\!CH_2 + B_2H_6 \xrightarrow{\text{THF}} 2(RCH_2CH_2)_3B$$

[6] J. D. Edwards, W. Gerrard, and M. F. Lappert, *J. Chem. Soc. (London)*, **1955**, 1470.

[7] H. C. Brown and P. A. Tierney, *J. Am. Chem. Soc.*, **80**, 1552 (1958).

[8] H. C. Brown and G. Zweifel, *J. Am. Chem. Soc.*, **83**, 1241 (1961).

Alternate procedures have been developed based on the amine-boranes. These are relatively stable materials prepared by the reaction of tertiary amines with diborane. Trimethylamine-borane[9] is a solid, m.p. 94°; pyridine-borane[10] and triethylamine-borane[11] are liquids. Olefins can be hydroborated by heating them with the amine-boranes at temperatures of 100 to 200°. Presumably the reaction involves an equilibrium dissociation of the amine-borane into diborane and amine, followed by a reaction of the olefin with the diborane.[12,13]

$$R_3N:BH_3 \rightleftharpoons R_3N + \tfrac{1}{2}(BH_3)_2$$

$$3RCH\!\!=\!\!CH_2 + \tfrac{1}{2}(BH_3)_2 \longrightarrow (RCH_2CH_2)_3B$$

Unfortunately, many private communications were received to the effect that the nonavailability of diglyme in countries abroad was a major handicap to the adoption of the hydroboration reaction as a laboratory procedure. For example, Dulou and Chrétien-Bessière reported that in order to circumvent this difficulty they generated diborane from the reaction of lithium aluminum hydride and boron trifluoride in ether solution[14] and passed the gas into an ether solution of the olefin.[15] Sondheimer and his co-workers avoided the external generation of diborane by adding lithium aluminum hydride to a solution of the olefin and boron trifluoride in ether solution.[16,17]

Since the original discovery of the alkali metal hydride route to diborane and the alkali metal borohydrides,[18] we have been exploring the interaction of various Lewis acids of boron with these hydrides and borohydrides (Chapter 4). It appeared desirable to utilize the available information to explore new hydroboration procedures, which would circumvent the requirements for any specific solvent, hydride source, or acid. Since iso-

[9] H. I. Schlesinger and A. B. Burg, *J. Am. Chem. Soc.*, **60**, 296 (1938).

[10] H. C. Brown, H. I. Schlesinger, and S. Z. Cardon, *J. Am. Chem. Soc.*, **64**, 325 (1942).

[11] A. C. Boyd, Ph. D. thesis, Purdue University Libraries.

[12] M. F. Hawthorne, *J. Org. Chem.*, **23**, 1788 (1958).

[13] E. C. Ashby, *J. Am. Chem. Soc.*, **81**, 4791 (1959).

[14] I. Shapiro, H. G. Weiss, M. Schmich, S. Skolnik, and G. B. L. Smith, *J. Am. Chem. Soc.*, **74**, 901 (1952).

[15] R. Dulou and Y. Chrétien-Bessière, *Bull. soc. chim. France*, **1959**, 1362.

[16] S. Wolfe, M. Nussim, Y. Mazur, and F. Sondheimer, *J. Org. Chem.*, **24**, 1034 (1959).

[17] F. Sondheimer and S. Wolfe, *Canadian J. Chem.*, **37**, 1870 (1959).

[18] H. I. Schlesinger, H. C. Brown, et al., *J. Am. Chem. Soc.*, **75**, 186 (1953).

merization of the organoborane becomes important at temperatures above 100° (Chapter 9), we emphasized reactions that would proceed at a reasonable rate at room temperature. Fortunately, it proved possible to realize these objectives.[19]

A simple standard procedure was followed. One hundred mmoles of 1-octene was mixed with 110 mmoles of "hydride" (10 per cent excess) in sufficient solvent to make the volume 70 cc. Through a dropping funnel was added 30 cc of a solution of the acid in a convenient solvent, generally ether, over a period of 1 hour, maintaining the temperature at 25°. At this point a sample was removed and analyzed for residual 1-octene by gas chromatography. The reaction mixture was again analyzed after standing 1 hour at 25°. If the reaction was still incomplete, it was heated to reflux for ethyl ether and tetrahydrofuran and to 75° for diglyme and analyzed after 1 hour and at regular intervals thereafter.

Lithium Borohydride

Lithium borohydride is readily soluble in ethyl ether, tetrahydrofuran, and diglyme. Consequently it was possible to examine the hydroboration reaction in all these solvents with a wide range of acids.

Boron trifluoride etherate achieves almost quantitative hydroboration in all solvents within 1 hour at 25°.

$$12RCH{=}CH_2 + 3LiBH_4 + 4BF_3 \longrightarrow 4(RCH_2CH_2)_3B + LiBF_4$$

It is known that in ethyl ether at 25° lithium borohydride will react with boron trifluoride in the ratio $3LiBH_4/BF_3$.[20] This reduced ratio is adequate in ethyl ether and tetrahydrofuran. However, in diglyme the reaction requires higher temperatures to achieve completion.

$$12RCH{=}CH_2 + 3LiBH_4 + BF_3 \longrightarrow 4(RCH_2CH_2)_3B + 3LiF$$

Boron trichloride does not form a complex chloroborate in ether solvents. Consequently, in this case there is no ambiguity as to the re-

[19] H. C. Brown, K. J. Murray, L. J. Murray, J. A. Snover, and G. Zweifel, *J. Am. Chem. Soc.*, **82**, 4233 (1960).

[20] J. R. Elliott, E. M. Boldebuck, and G. F. Roedel, *J. Am. Chem. Soc.*, **74**, 5047 (1952).

quired stoichiometry. The disadvantages of handling boron trichloride were discussed earlier.

$$12RCH{=}CH_2 + 3LiBH_4 + BCl_3 \longrightarrow 4(RCH_2CH_2)_3B + 3LiCl$$

A mixture of aluminum chloride and methyl borate behaves in many of these reactions as the equivalent of boron trichloride.[21] In ethyl ether a 77 per cent yield is realized, whereas in tetrahydrofuran and diglyme yields of better than 90 per cent are realized after 1 hour of heating.

$$12RCH{=}CH_2 + 3LiBH_4 + B(OCH_3)_3 + AlCl_3 \longrightarrow$$
$$4(RCH_2CH_2)_3B + Al(OCH_3)_3 + 3LiCl$$

The utility of aluminum chloride has been discussed. Titanium tetrachloride also proved to be effective. Unfortunately, in this case the reaction is accompanied by the formation of dark solids which interfere with the hydrogen peroxide oxidation.

The use of either hydrogen chloride or sulfuric acid dissolved in ether provides an alternative convenient means of achieving hydroboration. Such solutions are readily prepared and represent a satisfactory procedure in cases where the loss of "hydride" is not important.

$$6RCH{=}CH_2 + 2LiBH_4 + H_2SO_4 \longrightarrow 2(RCH_2CH_2)_3B + 2H_2 + Li_2SO_4$$

Sodium Borohydride

The fact that, of the standard solvents, sodium borohydride is soluble only in diglyme considerably limits the possibilities. However, in this solvent all of the acids are quite effective: boron trifluoride etherate, boron trichloride, aluminum chloride–methyl borate, titanium tetrachloride, hydrogen chloride, and sulfuric acid.

We attempted to utilize suspensions of sodium borohydride in ethyl ether and tetrahydrofuran with boron trifluoride etherate. In ethyl ether the reaction is very sluggish, but it proceeds satisfactorily in tetrahydrofuran, giving 72 per cent reaction after 1 hour at room temperature and 98 per cent reaction in 6 hours. If the boron trifluoride etherate is added to the reaction mixture as a solution in tetrahydrofuran, rather than in

[21] H. C. Brown and B. C. Subba Rao, unpublished observations.

ethyl ether, the reaction is very satisfactory, 99 per cent complete in 1 hour at 25°.

$$12RCH\!\!=\!\!CH_2 + 3NaBH_4 + 4BF_3\!:\!O(C_2H_5)_2 \xrightarrow[25°]{THF}$$

$$4(RCH_2CH_2)_3B + 3NaBF_4 \downarrow + 4(C_2H_5)_2O$$

This procedure is a simple one which avoids the use of the solvent diglyme. The sodium fluoroborate is insoluble in tetrahydrofuran, so that the solution of the organoborane may be readily obtained free of the salt.

A suspension of sodium borohydride in tetrahydrofuran reacts smoothly with boron trifluoride etherate to provide a solution of diborane. Addition of the olefin to this solution provides an alternative convenient route to the organoborane.

$$3NaBH_4 + 4BF_3\!:\!O(C_2H_5)_2 \xrightarrow{THF} 4BH_3(soln) + 3NaBF_4 \downarrow + 4(C_2H_5)_2O$$

$$12RCH\!\!=\!\!CH_2 + 4BH_3 \longrightarrow 4(RCH_2CH_2)_3B$$

It appeared desirable to explore further the utilization of sodium borohydride in ethyl ether. We had previously noted that anhydrous zinc chloride reacts with sodium borohydride suspended in ethyl ether to provide an ether solution of zinc borohydride.[22] The addition of olefin and boron trifluoride etherate to this solution achieves an almost quantitative reaction.

$$2NaBH_4 + ZnCl_2 \xrightarrow{EE} Zn(BH_4)_2 + 2NaCl \downarrow$$

$$24RCH\!\!=\!\!CH_2 + 3Zn(BH_4)_2 + 8BF_3\!:\!O(C_2H_5)_2 \longrightarrow$$

$$8(RCH_2CH_2)_3B + 3Zn(BF_4)_2 + 8(C_2H_5)_2O$$

Finally, it proved possible to achieve the hydroboration in ethyl ether utilizing only catalytic quantities of zinc chloride.

$$12RCH\!\!=\!\!CH_2 + 3NaBH_4 + 4BF_3\!:\!O(C_2H_5)_2 \xrightarrow[EE,\ 25°]{10\%\ ZnCl_2}$$

$$4(RCH_2CH_2)_3B + 3NaBF_4 + 4(C_2H_5)_2O$$

[22] H. C. Brown and J. A. Snover, unpublished observations.

Consequently, it is now possible to utilize sodium borohydride for hydroborations in all three of the standard solvents investigated — diglyme, tetrahydrofuran, and ethyl ether.

Potassium Borohydride

Potassium borohydride is essentially insoluble in diglyme, tetrahydrofuran, and ethyl ether. No significant reaction is observed between boron trifluoride and potassium borohydride in ethyl ether or in tetrahydrofuran. Even in diglyme the reaction is sluggish — only 39 per cent conversion after 1 hour at room temperature. Fortunately, the reaction is satisfactory in triglyme, so this solvent makes it possible to utilize the potassium salt for hydroboration.

Potassium borohydride undergoes an exchange with lithium chloride in tetrahydrofuran suspension.[23] The solution can be utilized for the hydroboration as described previously for lithium borohydride. Finally, it appears possible to achieve the hydroboration in ethyl ether by utilizing zinc chloride.[24]

Lithium Aluminum Hydride

Sondheimer and his co-workers have demonstrated that 1-octene and similar olefins can be hydroborated conveniently by a mixture of lithium aluminum hydride and boron trifluoride etherate in ethyl ether.[16,17] We have also observed that boron trichloride or a 1 : 1 molar mixture of methyl borate and aluminum chloride serve to achieve hydroborations with this hydride.

One minor disadvantage of this system should be mentioned. The use of lithium aluminum hydride results in the formation of a voluminous precipitate of aluminum hydroxide, which complicates somewhat the isolation of the organoborane or the oxidation products.

Alkali Metal Hydrides

Sodium hydride reacts with boron trifluoride etherate to produce diborane.[7,18] In diglyme solution, in the presence of 1-octene, reaction readily proceeds to give a 99 per cent conversion of the olefin.

$$6RCH{=}CH_2 + 6NaH + 8BF_3 \colon O(C_2H_5)_2 \longrightarrow$$
$$2(RCH_2CH_2)_3B + 6NaBF_4 + 8(C_2H_5)_2O$$

[23] R. Paul and N. Joseph, *Bull. soc. chim. France*, **1952**, 550.
[24] R. Paul and N. Joseph, *Bull. soc. chim. France*, **1953**, 758.

In tetrahydrofuran the reaction likewise proceeds to completion in 1 hour at room temperature. However, in ethyl ether the reaction is sluggish, with only 18 per cent reaction indicated under these conditions.

Lithium hydride reacts readily in all three solvents. However, the use of boron trifluoride etherate in the ratio $3LiH/BF_3$ results in only 20 to 25 per cent reaction. Use of sufficient boron trifluoride to form lithium fluorborate, $LiBF_4$, resulted in essentially complete conversions.

Amine-Boranes

Hydroborations based on the use of pyridine-borane,[12] trimethylamine-borane,[13] or triethylamine-borane[13] require temperatures in the range 100 to 200°. It appeared possible that the application of Lewis acids to these materials might lower the reaction temperature to the desired range. It has been reported that diborane can be prepared by treatment of the amine-borane with boron trifluoride.[25,26]

Unfortunately, it turned out that the liberation of diborane from these addition compounds by boron trifluoride etherate is by no means easy. Thus treatment of trimethylamine-borane in ether solution with boron trifluoride does not result in the liberation of diborane. Similarly, treatment of a solution of 1-octene and trimethylamine-borane in ether with boron trifluoride etherate resulted in no detectable reaction of the 1-octene after 2 hours at 25°. After 24 hours under reflux (46°), there was observed only 15 per cent reaction. Similar results were obtained with pyridine-borane. The use of boron trichloride or aluminum trichloride did not alter the situation.

It appeared possible that the ether solvent was inhibiting the reaction. Therefore benzene was adopted as the solvent and the reaction of the amine-boranes with 1-octene examined in this solvent. At 25° there appeared to be approximately 5 per cent reaction in 6 hours between pyridine-borane and 1-octene under the influence of boron trifluoride-etherate. At 75° there was observed 50 per cent reaction in 6 hours, 90 per cent reaction in 16 hours. The reaction course is indicated in the following equation:

$$3RCH{=\!\!=}CH_2 + C_5H_5N\!:\!BH_3 + BF_3\!:\!O(C_2H_5)_2 \longrightarrow$$

$$(RCH_2CH_2)_3B + C_5H_5N\!:\!BF_3 + (C_2H_5)_2O$$

Under the same conditions, but in the absence of the boron trifluoride etherate, the reaction proceeded to 10 per cent of completion in 6 hours,

[25] G. W. Schaeffer and G. D. Barbaras, U.S. Patent 2,533,696 (Dec. 12, 1950).
[26] R. Köster and K. Ziegler, *Angew. Chem.*, **69**, 94 (1957).

40 per cent completion in 16 hours. Doubling the initial concentration of the pyridine-borane results in an increased conversion of the olefin in the same reaction time.

Trimethylamine-borane appears to be considerably less reactive than pyridine-borane, with only 15 per cent conversion of the 1-octene observed after 16 hours at 75°, as compared to 90 per cent for pyridine-borane under the same conditions.

From these results, it appears that the amine-boranes are less convenient than metal hydrides and double hydrides for the hydroboration of olefins under mild conditions.

Hawthorne has circumvented these difficulties by the synthesis of a new reagent, trimethylamine-t-butylborane.[27] This reagent was obtained by treating t-butylboroxine (from t-butylmagnesium chloride and methyl borate at low temperatures) with lithium aluminum hydride in the presence of trimethylamine. It achieves the hydroboration of olefins, as well as dienes and acetylenes, at a reasonable rate, 2 hours at 60°.[28,29]

The combined inductive and steric requirements of the t-butyl groups would be expected to result in an enhanced tendency for dissociation. The more favorable rate is presumably a reflection of the increased dissociation at low temperatures.

$$t\text{-}C_4H_9BH_2 \text{:} N(CH_3)_3 \rightleftharpoons t\text{-}C_4H_9BH_2 + N(CH_3)_3$$

$$t\text{-}C_4H_9BH_2 + 2RCH\!=\!CH_2 \longrightarrow t\text{-}C_4H_9B(CH_2CH_2R)_2$$

The chemistry of a related amine-borane (trimethylamine-thexyl-borane[30]), available via the hydroboration reaction, will be discussed in Chapter 12.

Conclusion

These studies have revealed that the essentially quantitative hydroboration of olefins can be achieved in a variety of solvents with a variety of reagents. It now becomes possible to select the procedure on the basis of the available reagents and the form in which the organoborane is preferred — for isolation or further reaction.

In Chapter 6 we shall review the available data on the full range of olefinic structures that can be accommodated in the hydroboration reaction.

[27] M. F. Hawthorne, *J. Am. Chem. Soc.*, **81**, 5836 (1959).
[28] M. F. Hawthorne, *J. Am. Chem. Soc.*, **82**, 748 (1960).
[29] M. F. Hawthorne, *J. Am. Chem. Soc.*, **83**, 2541 (1961).
[30] H. C. Brown and G. J. Klender, *J. Inorg. Chem.*, in press.

6 | Scope

The low-temperature hydroboration of olefins was first announced in 1956.[1] Since that date the hydroboration reaction has been applied to a large variety of olefinic structures, varying from simple alkenes to complex steroids and bicyclic terpenes. It is the purpose of this chapter to review the structural types that have been tested experimentally in order to delineate the scope of this reaction. In this discussion we shall consider only the reactions that proceed readily at 0 to 25°, and not such reactions as might occur under more vigorous, forcing conditions.

The application of the reaction to dienes and acetylenes will be considered in Chapters 15 and 16. Its application to unsaturated structures containing functional groups will be briefly considered, but a detailed discussion of the special problems encountered in this area will be presented in Chapter 19.

Monosubstituted Terminal Olefins, $RCH{=}CH_2$

Simple terminal olefins, from ethylene through 1-dodecene, undergo hydroboration readily to form the corresponding tri-n-alkylboranes. In other words, mere lengthening of the carbon chain appears to have no significant effect upon the ease with which the reaction proceeds.[2]

$$
\left.\begin{array}{l}
CH_3CH{=}CH_2 \\
CH_3(CH_2)_4CH{=}CH_2 \\
CH_3(CH_2)_9CH{=}CH_2
\end{array}\right\} \xrightarrow{\ HB\ } [CH_3(CH_2)_nCH_2CH_2]_3B
$$

Similarly, the reaction readily accommodates severe branching of the alkyl groups, as revealed by the series 1-butene, 3-methyl-1-butene, and 3,3-dimethyl-1-butene.

[1] H. C. Brown and B. C. Subba Rao, *J. Am. Chem. Soc.*, **78**, 5694 (1956).
[2] H. C. Brown and B. C. Subba Rao, *J. Am. Chem. Soc.*, **81**, 6428 (1959).

$$CH_3CH_2CH{=}CH_2 \qquad CH_3\overset{\underset{\displaystyle CH_3}{|}}{C}HCH{=}CH_2 \qquad CH_3\overset{\underset{\displaystyle CH_3}{|}}{\overset{\overset{\displaystyle CH_3}{|}}{C}}CH{=}CH_2$$

Actually, no difference has been observed in the ease with which these branched-chain derivatives undergo hydroboration as compared to the straight-chain 1-alkenes.

Disubstituted Terminal Olefins, $R_2C{=}CH_2$

The presence of two alkyl substituents in olefins of the type $R_2C{=}CH_2$ must increase considerably the steric requirements. Nevertheless, numerous derivatives of this kind have been converted to the trialkylboranes without evidence of any difficulty.

Among the reactants that have been converted to the organoboranes may be mentioned 2-methylpropene,[3] 2-methyl-1-butene,[4] 2-methyl-1-pentene,[5] and 2,4,4-trimethyl-1-pentene.[2]

$$CH_3CH_2CH_2\overset{\overset{\displaystyle CH_3}{|}}{C}{=}CH_2 \qquad CH_3\overset{\underset{\displaystyle CH_3}{|}}{\overset{\overset{\displaystyle CH_3}{|}}{C}}CH_2\overset{\overset{\displaystyle CH_3}{|}}{C}{=}CH_2$$

A particularly interesting series is 2-methylpropene, 2,3,3-trimethyl-1-butene,[6] and 1,1-di-t-butylethylene.[7]

[3] H. C. Brown and C. Snyder, *J. Am. Chem. Soc.*, **83**, 1001 (1961).

[4] H. C. Brown and G. Zweifel, *J. Am. Chem. Soc.*, **82**, 1504 (1960).

[5] H. C. Brown, K. J. Murray, L. J. Murray, J. A. Snover, and G. Zweifel, *J. Am. Chem. Soc.*, **82**, 4233 (1960).

[6] H. C. Brown, G. Zweifel, and N. R. Ayyanger, unpublished observations.

[7] M. S. Newman, A. Arkell, and T. Fukunaga, *J. Am. Chem. Soc.*, **82**, 2498 (1960).

It is gratifying that a structure as sterically hindered as 1,1-di-*t*-butylethylene must be still undergoes the hydroboration reaction with ease. It would be of considerable interest to know whether the hydroboration of this very bulky olefin proceeds to the trialkylborane stage or whether it stops at a lower reaction stage.

These highly branched structures are strongly susceptible to carbon structure rearrangements in many reactions involving attack on the double bond. However, in none of these cases has any evidence been observed for such rearrangements during hydroboration.

Disubstituted Internal Olefins, RCH=CHR

Simple internal olefins, such as 2-butene and 2-pentene,[2] undergo rapid conversion to the trialkylborane stage, with no evidence for any tendency to halt at an intermediate stage.

However, the more highly branched olefin, *trans*-4,4-dimethyl-2-pentene, undergoes rapid hydroboration only to the dialkylborane stage, with further reaction to the trialkylborane being relatively slow.[8] Similarly, *trans*-2,2,4,4-tetramethyl-3-hexene (*trans*-di-*t*-butylethylene) undergoes hydroboration only to the monoalkylborane stage.[9]

[8] H. C. Brown and G. Zweifel, *J. Am. Chem. Soc.*, **82**, 4708 (1960).
[9] T. J. Logan and T. J. Flautt, *J. Am. Chem. Soc.*, **82**, 3446 (1960).

Tri- and Tetrasubstituted Olefins, R_2C=CHR and R_2C=CR$_2$

2-Methyl-2-butene undergoes rapid hydroboration to the dialkylborane stage. Further reaction to the trialkylborane stage occurs, but is quite slow compared to the rate of the initial reaction.[10,11] The trisubstituted olefin, 2,4,4-trimethyl-2-pentene, exhibits a rapid reaction to the monoalkylborane stage, with a relatively slow further reaction to the dialkylborane.[11]

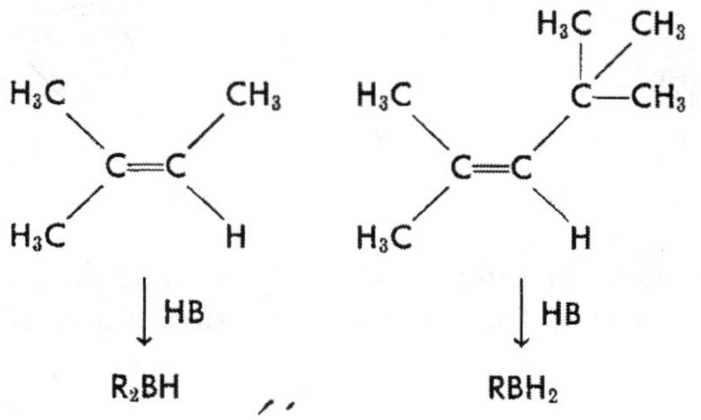

The tetrasubstituted olefin, 2,3-dimethyl-2-butene (tetramethylethylene), undergoes rapid hydroboration to the monoalkylborane stage,[2] with further reaction to the dialkylborane proceeding at a much lower rate.[11] Consequently, it is possible to summarize the behavior of the simple methyl substituted ethylenes as follows:

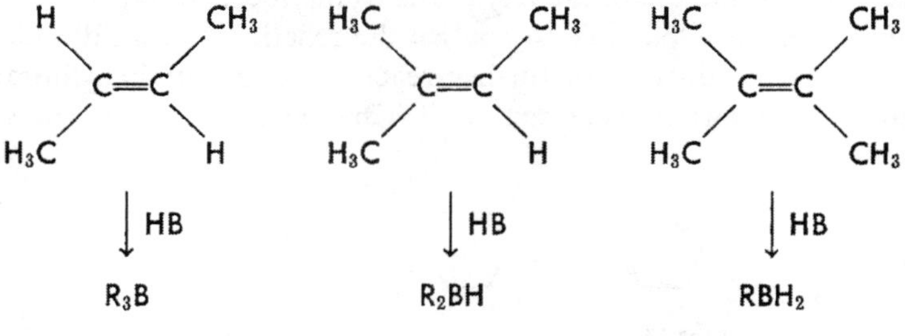

Phenyl-Substituted Ethylenes

The presence of aromatic substituents appears to offer no difficulty. Derivatives such as styrene, 1,1-diphenylethylene, *trans*-stilbene, and triphenylethylene undergo hydroboration readily.[2] Indeed, the steric requirements of the phenyl group appear to be considerably less than that

[10] H. C. Brown and G. Zweifel, *J. Am. Chem. Soc.*, **83**, 1241 (1961).

[11] H. C. Brown and A. W. Moerikofer, *J. Am. Chem. Soc.*, in press.

of *t*-butyl, so that the behavior of these derivatives resembles more closely that of the methyl olefins described above than of the corresponding structures containing *t*-butyl substituents.

It is unfortunate that the insolubility of tetraphenylethylene in the hydroboration solvents has prevented an examination of its behavior in this reaction.

Simple Cyclic Olefins, $(CH_2)_n$

The hydroboration of cyclopentene,[2] cyclohexene,[2] cycloheptene,[12] and cyclooctene[13] has been accomplished. With one exception the reactions proceed rapidly and quantitatively to the trialkylborane stage. In the case of cyclohexene, it is possible to control the reaction so that dicyclohexylborane is the product, with further reaction to tricyclohexylborane occurring at a somewhat decreased rate[11] (Chapter 11).

[12] H. C. Brown and A. W. Moerikofer, *J. Am. Chem. Soc.*, **83**, 3417 (1961).
[13] H. C. Brown and C. P. Garg, unpublished observations.

This exceptional behavior of cyclohexene appears to have its origin in two special characteristics of the cyclohexene system. First, cyclohexene is considerably less reactive toward hydroboration than the other cyclic olefins listed. Second, dicyclohexylborane (actually *sym*-tetracyclohexyldiborane) is a crystalline, relatively high melting solid which is insoluble and precipitates out of the hydroboration solvent.

The presence of the strain in the medium rings appears to enhance greatly the reactivity of these cyclic olefins in the hydroboration reaction (Chapter 13).

Substituted Cyclic Olefins, $(CH_2)_n$ $\begin{matrix} C-R \\ \| \\ C-H \end{matrix}$ **and** $(CH_2)_n$ $\begin{matrix} C-R \\ \| \\ C-R \end{matrix}$

The behavior of the 1-alkylcycloalkenes appears to parallel the behavior of the corresponding ethylene derivatives. Thus the hydroboration of 1-methylcyclohexene proceeds cleanly to the dialkylborane stage.[14] In the case of 1-methylcyclopentene the hydroboration also proceeds rapidly to the dialkylborane stage. However, the greater reactivity of the five- as compared to the six-membered-ring olefin is reflected in an enhanced tendency for the reaction to proceed further to the trialkylborane stage.[11]

Indeed, in caryophyllene the nine-membered ring exhibits such extraordinary reactivity that the hindered organoborane from trimethylethylene, disiamylborane, reacts preferentially with the methylcyclononene structure, in preference to the normally more reactive exocyclic methylene double bond.[15]

The middle-ring olefins exhibit other unusual characteristics. The hydroboration of 1-methylcyclooctene appeared to proceed normally, but the usual oxidation produced more than one alcohol. It appears that this system may be unusually susceptible to isomerization of the organoboron intermediate (Chapter 9).

The hydroboration of 1-phenylcyclohexene[14] appears to proceed to the dialkylborane stage, albeit relatively slowly, whereas 1-*t*-butylcyclohexene[16] appears to go to the monoalkylborane stage primarily.

[14] H. C. Brown and G. Zweifel, *J. Am. Chem. Soc.*, **83**, 2544 (1961).
[15] H. C. Brown and K. P. Singh, unpublished observations.
[16] H. C. Brown and G. Zweifel, unpublished observations.

The hydroboration of a number of trisubstituted olefins in the decalyl series has been described by Sondheimer and Wolfe.[17]   However, the authors utilized a large excess of reagent, so nothing can be stated as to the stoichiometry of the reaction.

The disubstituted cycloalkenes, 1,2-dimethylcyclopentene and 1,2-dimethylcyclohexene, behave similarly to tetramethylethylene.[14]   In all cases the reaction proceeds rapidly to the monoalkylborane stage and only slowly beyond.

Bicyclic Olefins

The hydroboration of β-pinene proceeds readily to the trialkylborane stage.[11,14]   Camphene also undergoes hydroboration readily, presumably to the trialkylborane stage, although the stoichiometry does not appear to have been definitely established.[18]

[17] F. Sondheimer and S. Wolfe, *Canadian J. Chem.*, **37**, 1870 (1959).

[18] R. Dulou and Y. Chrétien-Bessière, *Bull. soc. chim. France*, **1959**, 1362.

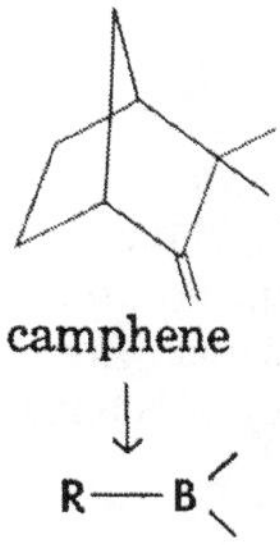

camphene

↓

R — B⟨

The internal double bond of norbornene reacts with great ease, the reaction proceeding rapidly to the trialkylborane stage.[14]

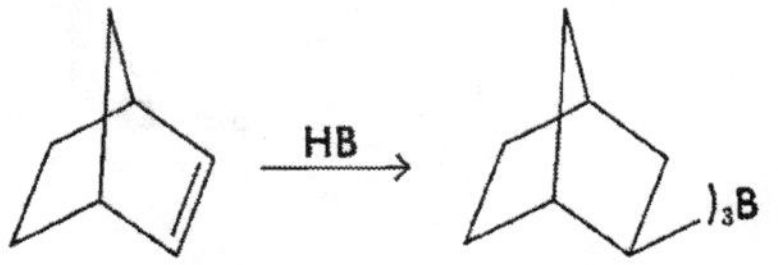

The presence of a methyl substituent on the double bond of the bicyclic system causes the reaction to proceed only to the intermediate dialkylborane stage. For example, the hydroboration of α-pinene proceeds to the formation of diisopinocampheylborane,[14] but there is some evidence of a tendency to stop somewhat short of this stage [19] (Chapters 11 and 12).

No difficulty appears to have been experienced in achieving the hydroboration of the more complex structures, exo-dihydrodicyclopentadiene,[20] and isodrin.[21,22]

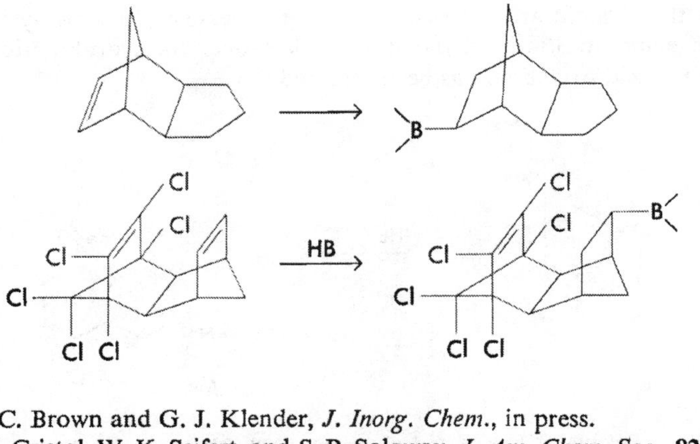

[19] H. C. Brown and G. J. Klender, J. Inorg. Chem., in press.
[20] S. J. Cristol, W. K. Seifert, and S. B. Soloway, J. Am. Chem. Soc., 82, 2351 (1960).
[21] R. C. Cookson and E. Crundwell, Chem. & Ind. (London), 1959, 703.
[22] P. Bruck, D. Thompson, and S. Winstein, Chem. & Ind. (London), 1960, 405.

Steroids

Numerous steroids have been subjected to hydroboration, beginning with Wechter's conversion of cholesterol into the corresponding diol.[23] Wechter noted that the hydroboration of cholesterol proceeds to the dialkylborane stage, in accordance with the behavior that is observed with the simple 1-alkylcyclohexenes of related structure.

Nussim and Sondheimer synthesized a number of 11-oxygenated steroids.[24]

In the alkaloid area, the conversion of conessine (3β-dimethylamino-con-5-enine) to 3β-dimethylaminoconanin-6-one, via hydroboration and chromic acid oxidation, has been reported.[25]

[23] W. J. Wechter, *Chem. & Ind. (London)*, **1959**, 294.
[24] M. Nussim and F. Sondheimer, *Chem. & Ind. (London)*, **1960**, 400.
[25] R. Pappo, *J. Am. Chem. Soc.*, **81**, 1010 (1959).

Two cases have been reported in which a molecule with a double bond resists the hydroboration reaction. It was noted by Wechter that 3,20-bis-cycloethylenedioxy-5β-pregn-9(11)-ene failed to undergo reaction under the usual hydroboration conditions (AlCl₃ + 3NaBH₄).[23]

Similarly, test-9(11)-ene with the same A/B-*cis* junction failed to react.[24]

However, Nussim and Sondheimer have observed that the analogous compounds with the A/B-*trans* junction undergo hydroboration without difficulty.[24]

Olefins Containing Substituent Groups

Groups that are relatively inert to the action of diborane, such as alkoxy or halo, appear to offer little or no difficulty. Thus the hydroboration of isodrin, with its many chlorine substituents, proceeds normally.[21,22] Similarly, the hydroboration of *p*-chlorostyrene and *p*-methoxystyrene proceeds to the corresponding trialkylborane stage, the only exceptional feature being a modest change in the orientation taken by the entering boron atom[8] (Chapter 8).

In the same way no difficulties are observed in the hydroboration of vinylethyl ether[26] or vinyltrimethylsilane.[27]

When the halogen is relatively near the double bond, difficulties are encountered. It appears that the structure X—C—C—B$\langle$ undergoes elimination easily. Consequently, the hydroboration of vinyl chloride was observed to yield complex mixtures.[28] On the other hand, allyl chloride undergoes satisfactory hydroboration to tri-(γ-chloropropyl)-borane.[28] Similarly, no difficulty was experienced in the hydroboration of allylmethylsulfide.[29]

Even where the molecules contain functional groups that are reducible by diborane (Chapter 17), the reaction of diborane with the double bond is so fast that conversion to the organoborane can be achieved in nearly quantitative yield. For example, the hydroboration of methyl oleate proceeds quite simply to provide, after oxidation, the 9- and 10-hydroxy-octadecanoic ester.[30]

Finally, it is of interest to point out that it has been proved possible to extend the hydroboration reaction to the sugar field.[31]

[26] B. M. Mikhailov and T. A. Shchegoleva, *Izvest. Akad. Nauk. SSSR*, **1959**, 546.

[27] D. Seyferth, *J. Inorg. & Nuclear Chem.*, **7**, 152 (1958).

[28] M. F. Hawthorne and J. A. Dupont, *J. Am. Chem. Soc.*, **80**, 5830 (1958).

[29] H. C. Brown and K. Murray, *J. Am. Chem. Soc.*, **81**, 4108 (1959).

[30] S. P. Fore and W. G. Bickford, *J. Org. Chem.*, **24**, 920 (1959).

[31] M. L. Wolfrom and T. E. Whiteley, Abstracts of Papers, 137th Am. Chem. Soc. Meeting, Sec. 2–D.

7 | Directive Effects

The original hydroboration experiments had indicated that the hydroboration reaction proceeded to place the boron atom predominantly at the less substituted of the two carbon atoms of the double bond.[1,2]

The organoborane is readily converted into the corresponding alcohol by oxidation with alkaline hydrogen peroxide. Consequently, hydroboration followed by oxidation provides a very valuable procedure for the anti-Markownikoff hydration of olefins.[3]

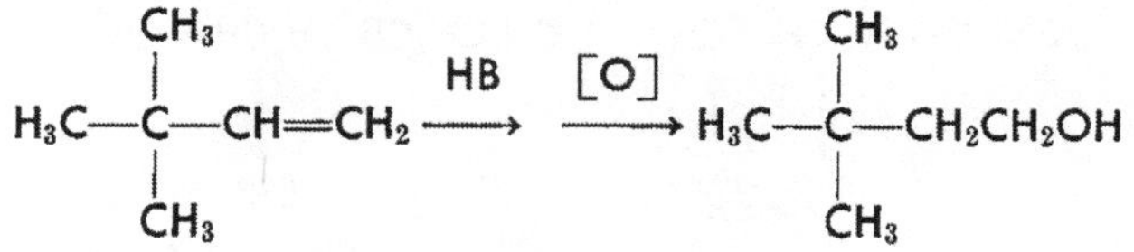

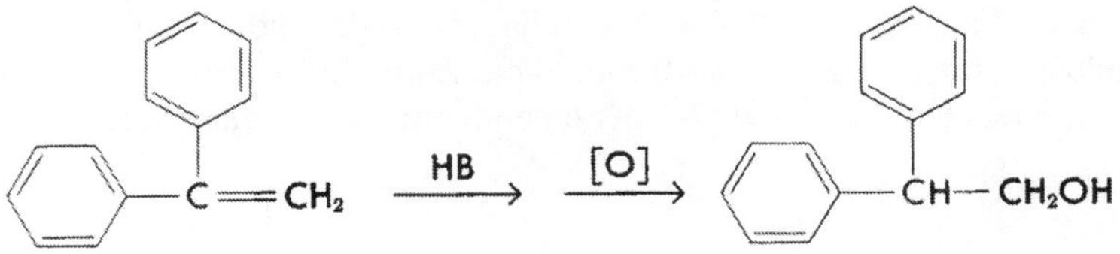

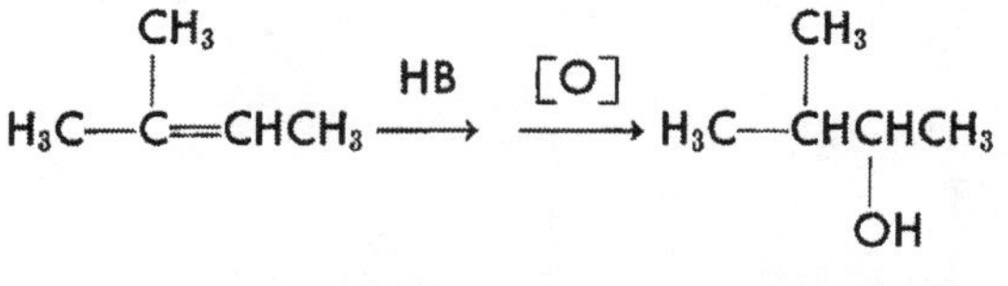

$$H_3C-\underset{\underset{CH_3}{|}}{\overset{\overset{CH_3}{|}}{C}}-CH=CH_2 \xrightarrow{HB} \xrightarrow{[O]} H_3C-\underset{\underset{CH_3}{|}}{\overset{\overset{CH_3}{|}}{C}}-CH_2CH_2OH$$

$$C_6H_5\underset{\underset{C_6H_5}{}}{C}=CH_2 \xrightarrow{HB} \xrightarrow{[O]} C_6H_5-\underset{\underset{C_6H_5}{}}{CH}-CH_2OH$$

$$H_3C-\underset{\underset{}{\overset{\overset{CH_3}{|}}{C}}}=CHCH_3 \xrightarrow{HB} \xrightarrow{[O]} H_3C-\underset{\underset{OH}{|}}{\overset{\overset{CH_3}{|}}{CH}}CHCH_3$$

---

[1] H. C. Brown and B. C. Subba Rao, *J. Am. Chem. Soc.*, **78**, 5694 (1956).

[2] H. C. Brown and B. C. Subba Rao, *J. Am. Chem. Soc.*, **81**, 6423 (1959).

[3] G. Zweifel and H. C. Brown, Hydration of Olefins, Dienes and Acetylenes Via Hydroboration, in "Organic Reactions," John Wiley & Sons, Inc., New York, in press.

113

The available evidence is that the hydrogen peroxide oxidation of organoboranes is a quantitative reaction, placing a hydroxyl group in the precise position previously occupied by the boron atom. Gas chromatography provides a powerful analytical tool for the analysis of mixtures of isomeric alcohols. Accordingly, a detailed study of directive effects in the hydroboration of olefins of various structural types was undertaken, utilizing the hydrogen peroxide oxidation of the products to establish the orientation.[4]

Monosubstituted Terminal Olefins, $RCH=CH_2$

Simple straight-chain terminal olefins, such as 1-butene, 1-pentene, and 1-hexene, give predominant addition of the boron atom to the terminal carbon atom (93 to 94 per cent). Only a minor amount of addition to give the secondary alkyl boron product is observed (6 to 7 per cent).

$$CH_3CH_2CH=CH_2 \qquad CH_3CH_2CH_2CH_2CH=CH_2$$

$$7\% \ \ 93\% \qquad\qquad 6\% \ \ 94\%$$

Branching of the alkyl group adjacent to the double bond does not influence the direction of addition. Thus, 3-methyl-1-butene, 3,3-dimethyl-1-butene, and 4,4-dimethyl-1-pentene give the same distribution.

$$7\% \ \ 93\% \qquad\quad 6\% \ \ 94\% \qquad\quad 6\% \ \ 94\%$$

The phenyl group modifies the distribution significantly, with 20 per cent of the boron becoming attached *alpha* to the phenyl group. The influence of the phenyl group is diminished but is still evident in allylbenzene with a 90:10 distribution.

[4] H. C. Brown and G. Zweifel, *J. Am. Chem. Soc.*, **82**, 4708 (1960).

$$\underset{20\%\quad 80\%}{\text{C}_6\text{H}_5-\overset{\uparrow}{\text{CH}}=\overset{\uparrow}{\text{CH}_2}} \qquad \underset{10\%\quad 90\%}{\text{C}_6\text{H}_5-\overset{\uparrow}{\text{CH}_2}\text{CH}=\overset{\uparrow}{\text{CH}_2}}$$

Styrene was hydroborated by a variety of techniques in several solvents (ethyl ether, tetrahydrofuran, and diglyme). No significant variation was observed in the isomer distribution.

Disubstituted Terminal Olefins, $R_2C{=}CH_2$

In the case of disubstituted terminal olefins, the directive effect of the two substituents is overwhelming and results in the almost complete addition of the boron atom to the terminal carbon atom. Thus 2-methyl-1-butene gives 99 per cent of the primary alkylborane with only 1 per cent of the tertiary derivative. Similar results are observed with 2,4,4-trimethyl-1-pentene and α-methylstyrene.

$$\underset{1\%\quad 99\%}{\text{CH}_3\text{CH}_2\overset{\overset{\textstyle CH_3}{|}}{\underset{\uparrow}{\text{C}}}=\overset{\uparrow}{\text{CH}_2}} \qquad \underset{\text{tr.}\quad 100\%}{C_6H_5-\overset{\overset{\textstyle CH_3}{|}}{\underset{\uparrow}{\text{C}}}=\overset{\uparrow}{\text{CH}_2}}$$

Disubstituted Internal Olefins, $RCH{=}CHR'$

2-Pentene and 2-hexene undergo hydroboration to place the boron in approximately equal amounts on the two carbon atoms. Only minor variations in the distribution are observed for the cis and trans isomers.

$$\underset{45\%\quad 55\%}{\underset{\text{CH}_3\text{CH}_2}{\nearrow}\overset{\uparrow}{\text{CH}}=\overset{\uparrow}{\text{CH}}\underset{\text{CH}_3}{\nwarrow}} \qquad \underset{49\%\quad 51\%}{\text{CH}_3\text{CH}_2\overset{\uparrow}{\text{CH}}=\overset{\uparrow}{\text{CH}}\underset{\text{CH}_3}{\nwarrow}}$$

The observation that the symmetrical olefin, cis-3-hexene, yields 100 per cent of 3-hexanol confirms the conclusion that isomerization does not occur under these conditions, so that this procedure accurately gives the distribution in the initial hydroboration product.

Branching of one of the two alkyl groups attached to the double bond, as in *trans*-4-methyl-2-pentene and *trans*-4,4-dimethyl-2-pentene, results in a small preference of the boron atom for that carbon atom of the double bond that is adjacent to the less-branched of the alkyl substituents.

In the same way, no significant directive effect was observed in the hydroboration of 3-methylcyclopentene, 3-methylcyclohexene, and 3,3-dimethylcyclohexene.[5]

Sondheimer and Nussim likewise observed no significant effect in the hydroboration of Δ^1-cholestene. They identified in the product 35 per cent of cholestan-1α-ol and 40 per cent of cholestan-2α-ol.[6]

[5] H. C. Brown and G. Zweifel, *J. Am. Chem. Soc.*, **83**, 2544 (1961).
[6] F. Sondheimer and M. Nussim, *J. Org. Chem.*, **26**, 630 (1961).

It should be mentioned that in certain of these cases it has been possible to achieve steric control of the direction of hydroboration by using the reagent disiamylborane (Chapter 13).

The styrene results discussed earlier indicate that the phenyl group is less effective than the alkyl group in directing the boron atom to the terminal carbon. The same relative effect of the phenyl and methyl groups is shown by the distribution observed for *trans*-1-phenylpropene — 85 per cent α- and 15 per cent β-.

Trisubstituted Olefins, $R_2C{=}CHR$

2-Methyl-2-butene undergoes hydroboration to give 98 per cent of the secondary derivative, with only 2 per cent of the tertiary compound. The same relative distribution between the second and tertiary positions is observed in 2,4,4-trimethyl-2-pentene.

In spite of the influence of the aryl group in directing substitution toward it, the hydroboration of *cis* and *trans*-2-*p*-anisyl-2-butenes occurs predominantly at the secondary position.[7]

[7] E. L. Allred, J. Sonnenberg, and S. Winstein, *J. Org. Chem.*, **25**, 26 (1960).

72% isolated

p-Substituted Styrenes

A p-methyl substituent in styrene exerts only a minor effect, decreasing the amount of *alpha* substitution from the 20 per cent observed for styrene to 18 per cent. A p-methoxy substituent exerts a far-more-powerful effect, decreasing the amount of *alpha* substitution to 9 per cent. Finally, the p-chloro substituent increases the yield of *alpha* product to 35 per cent.

Theoretical Considerations

The addition of diborane to olefins must involve at least three distinct stages, in which one, two, and three carbon-boron bonds are formed.

$$\underset{|}{\overset{|}{C}}=\underset{|}{\overset{|}{C}} + BH_3 \longrightarrow H-\underset{|}{\overset{|}{C}}-\underset{|}{\overset{|}{C}}-BH_2$$

$$\underset{|}{\overset{|}{C}}=\underset{|}{\overset{|}{C}} + RBH_2 \longrightarrow H-\underset{|}{\overset{|}{C}}-\underset{|}{\overset{|}{C}}-BHR$$

$$\underset{|}{\overset{|}{C}}=\underset{|}{\overset{|}{C}} + R_2BH \longrightarrow H-\underset{|}{\overset{|}{C}}-\underset{|}{\overset{|}{C}}-BR_2$$

Actually, if we consider the probability that the reaction may involve the dimeric species, even more steps may be involved.

The direction of addition was established by oxidizing the final product to the corresponding alcohols. Clearly, the final isomer distribution is the result of at least three successive stages, and there is no assurance that the distribution of the boron between the two available carbon atoms of the double bond is the same in each stage. Indeed, we have evidence that the distribution can vary with successive substitution of the borane group.

Fortunately, this does not constitute a major problem in considering the data. With yields of 94 to 98 per cent of a single isomer realized, it is apparent that the direction of addition cannot vary greatly in the individual stages of these particular reactions. Only in the case of additions to styrene (20 per cent α-) and p-chlorostyrene (35 per cent α-) is there uncertainty as to whether the boron atom prefers the α- or β-carbon atom in the initial stage of the addition reaction.

In Chapter 6 the available data on the stoichiometry of the hydroboration reaction were presented. The hydroboration of *cis-* and *trans-2-*butene proceeds cleanly to the trialkylborane stage, whereas hydroboration of the related olefin, *trans-t-*butylmethylethylene, proceeds readily only to the dialkylborane stage,[4] and that of *trans-*di-*t-*butylethylene halts at the monoborane stage.[8]

These results clearly point to the importance of steric effects in controlling the extent of reaction between a given olefin and diborane.

That steric effects also play a role in controlling the direction of hydroboration is indicated by the results realized with disiamylborane,

[8] T. J. Logan and T. J. Flautt, *J. Am. Chem. Soc.*, **82**, 3446 (1960).

formed in the reaction of diborane with 2-methyl-2-butene (Chapter 13). Whereas the reaction of diborane with 1-hexene yields 6 per cent of the secondary alkyl boron derivative, the hindered dialkylborane yields a product with 1-hexene which contains no more than 1 per cent of the boron at the secondary position.[9] Consequently, it is tempting to ascribe the observed preference of the boron atom for the less substituted of the two carbon atoms constituting the double bond to the operation of steric effects.

It would be possible on this basis to account for the high preference of the boron atom for the terminal carbon atom of monoalkylethylenes, $RCH=CH_2$ (94 per cent), the terminal atom of 1,1-dialkylethylenes, $R_2C=CH_2$ (99 per cent), and the secondary carbon atom of trialkylethylenes, $R_2C=CHR$ (98 per cent).

However, the data reveal that steric effects cannot be the complete answer. No significant change in the direction of addition is observed for "ethylethylene" (93 per cent 1-), isopropylethylene (94 per cent 1-), *t*-butylethylene (94 per cent 1-), or neopentylethylene (93 per cent 1-). Moreover, the shifts in distribution observed for $RCH=CHCH_3$ (51 per cent 2- for R = ethyl, 57 per cent 2- for R = isopropyl, 58 per cent 2- for R = *t*-butyl), although in the direction to be predicted for steric control, appear far too small to account for the major directive effects observed in the terminal and trisubstituted olefins.

Also arguing against a steric basis for the observed directive influence is the vast difference in orientation observed in *t*-butylethylene (94 per cent 1-) and vinyltrimethylsilane (63 per cent 1-).[10]

Finally, the relatively large amount of addition in the position *alpha* to the aromatic ring in styrene (20 per cent) and *trans*-phenylmethylethylene (85 per cent) does not appear to be explicable in terms of steric influence alone.

[9] H. C. Brown and G. Zweifel, *J. Am. Chem. Soc.*, **83**, 1241 (1961).
[10] D. Seyferth, *J. Inorg. & Nuclear Chem.*, **7**, 152 (1958).

Attention should be called to the similar results realized in the addition of diethylaluminum hydride to 1-hexene[11] and to styrene.[12] The 1-hexene adds the aluminum-hydrogen bond to place the aluminum practically exclusively on the terminal carbon atom, whereas in the case of styrene 22 to 24 per cent of the aluminum becomes attached to the alpha carbon atom of the side chain.

The available data clearly require that electronic influences play an important role in controlling the direction of addition of the boron-hydrogen bond to the double bond. This conclusion is confirmed by the marked influence exerted by *para* substituents in the styrene molecule on the direction of addition.

The addition of diborane to cyclic olefins[5] and to acetylenes[13] occurs cleanly *cis*. Consequently, it appears that the addition must involve a four-center transition state (Chapter 8).

$$R-\underset{\underset{H}{|}}{C}=\underset{\underset{H}{|}}{C}-H + {>}B-H \longrightarrow R-\underset{\underset{H---B---}{|}}{C}=\underset{\underset{H}{|}}{C}-H \longrightarrow RCH_2CH_2\underset{\underset{\wedge}{|}}{B}$$

The boron-hydrogen bond is presumably polarized, with the hydrogen having some hydridic character. Consequently, the electronic shifts generally used to account for the normal addition of hydrogen chloride to propylene[14] can be applied to account for the preferred addition of the boron atom to the terminal position.

$$H_2C-CH=CH_2 + {>}B-H \longrightarrow H_2C-\overset{\delta+}{CH}=\overset{\delta-}{CH_2}\underset{\underset{\delta-}{H}---\underset{|\delta+}{B}-}{}$$

[11] K. Ziegler, H. G. Gellert, H. Martin, K. Nagel, and J. Schneider, *Ann.*, **589**, 91 (1954).

[12] G. Natta, P. Pino, G. Mazzanti, P. Longi, and F. Bernardini, *J. Am. Chem. Soc.*, **81**, 2561 (1959).

[13] H. C. Brown and G. Zweifel, *J. Am. Chem. Soc.*, **81**, 1512 (1959).

[14] G. W. Wheland, "Resonance in Organic Chemistry," John Wiley & Sons, Inc., New York, 1955, pp. 430-442.

The same explanation can be extended to account for the enhanced prefer-ence of boron for the terminal carbon atom of the 2,2-dialkylethylenes (99 per cent 1-), as well as for the preferred substitution in the secondary position (98 per cent) of the trisubstituted olefins.

In the case of styrene, similar arguments can be used to account for the observed preference of the boron atom for the terminal position.

In order to account for the enhanced substitution in the *alpha* position of this molecule, we must recognize that the phenyl group can not only stabilize a positive charge in the *alpha* position by supplying electron density, it can also stabilize a negative charge in the *alpha* position by absorbing electron density.   This effect is apparent in the relatively high stability of benzylic anions.[14]

This transition state would be stabilized by an electron-withdrawing substituent, such as *p*-chloro-, and rendered less stable by an electron-supplying substituent, such as *p*-methoxy.

On this basis it appears that the observed anti-Markownikoff addition is primarily the result of the hydridic polarization of the boron-hydrogen bond.

8 | Stereochemistry

The hydroboration of substituted cyclic olefins, such as the 1-alkylcyclo-alkenes, offered the possibility of establishing the stereochemical aspects of the hydroboration reaction. Accordingly, a detailed study was made of the hydroboration of the compounds 1-methylcyclopentene and 1-methyl-cyclohexene.

It was established that in these olefins the boron atom becomes attached to the 2 position of the ring, as predicted by the directive effects results (Chapter 7). Unfortunately, there is no simple method available for establishing the stereochemistry of the boron intermediate. However, oxidation with alkaline hydrogen peroxide yields the pure *trans*-2-methylcycloal-kanols of established structures. *It follows that the over-all hydration proceeds as an anti-Markownikoff, cis hydration of the double bond.*[1]

Shortly afterward, Wechter reported that the related hydroboration-oxidation of cholesterol likewise involves a pure *cis* hydration of the double bond.[2]

[1] H. C. Brown and G. Zweifel, *J. Am. Chem. Soc.*, **81**, 247 (1959).
[2] W. J. Wechter, *Chem. & Ind. (London)*, **1959**, 294.

Obviously, such a simple, stereochemically defined hydration of double bonds has exceedingly important implications for synthetic chemistry, and a detailed study of the reaction was therefore undertaken to establish the full scope of this *cis* hydration and to unravel the mechanism.[3]

Monocyclic Systems

The hydroboration of 2-methyl-2-butene proceeds rapidly to the dialkylborane stage.   1-Methylcyclopentene and 1-methylcyclohexene are also trisubstituted olefins.   Their hydroboration likewise proceeds rapidly to the dialkylborane stage, with further reaction being quite slow.   Oxidation with alkaline hydrogen peroxide yields the 2-methylcycloalkanols.

Gas-chromatographic examination of the product from 1-methylcyclopentene reveals the presence of 1.5 per cent of 1-methylcyclopentanol and 98.5 per cent of *trans*-2-methylcyclopentanol.   No evidence for the presence of the *cis* isomer has been observed.   In the same way, examination of the crude product from 1-methylcyclohexene reveals the presence of 1.5 per cent of 1-methylcyclohexanol, 97.2 per cent of *trans*-2-methylcyclohexanol, and 0.8 per cent of a minor component.   This minor component might be the *cis* isomer, but definite identification was not possible.

[3] H. C. Brown and G. Zweifel, *J. Am. Chem. Soc.*, **83**, 2544 (1961).

1-Methylcyclooctene yields anomalous results. Examination of the product reveals the presence of three components. The results are compatible with an unusually facile isomerization of the 2-methylcyclooctyl-borane derivative presumably formed in the hydroboration stage.

Application of the hydroboration-oxidation procedure to 1-phenyl-cyclohexene proceeds normally. *Trans*-2-phenylcyclohexanol is obtained in 79 per cent yield (isolated).

These experiments establish that the hydroboration of 1-methylcyclo-pentene, 1-methylcyclohexene, and 1-phenylcyclohexene, followed by oxidation with alkaline hydrogen peroxide, proceeds exceedingly cleanly stereochemically, to give the pure *trans*-2-methylcyclopentanol, *trans*-2-methylcyclohexanol, and *trans*-2-phenylcyclohexanol, corresponding to an essentially quantitative *anti*-Markownikoff *cis* addition of water to the double bond.

CH₃ —HB→ [O]→ (H₃C, H, OH) (cyclopentene → trans-2-methylcyclopentanol)

CH₃ —HB→ [O]→ (H₃C, H, OH) (cyclohexene → trans-2-methylcyclohexanol)

(phenyl) —HB→ [O]→ (phenyl, H, OH) (1-phenylcyclohexene → trans-2-phenylcyclohexanol)

In each of these cases, the alcohol obtained is the more stable of the two possible isomeric products. Accordingly, it appeared desirable to examine the hydroboration of 1,2-dimethylcyclopentene and 1,2-dimethylcyclo-hexene. In these cases *cis* hydration would result in the formation of the thermodynamically less stable isomer.

The hydroboration of 1,2-dimethylcyclopentene and 1,2-dimethylcyclo-hexene proceeds readily only to the monoalkylborane stage, as previously noted for the related acyclic olefin, tetramethylethylene.   Oxidation produces pure *cis*-1,2-dimethylcyclopentanol and *cis*-1,2-dimethylcyclohexanol.

Consequently, in these cases also the procedure brings about a pure *cis* hydration of the double bond, even though the products are presumably the thermodynamically less stable of the two possible isomers. These results argue against the possibility that the stereochemical course of the reaction is controlled by product stability.

Finally, the remarkable stereochemical specificity of the reaction provides a further argument that the reaction course is determined by the mechanism and not by minor differences in the thermodynamic stability of the products.

Bicyclic Systems

The hydroboration of norbornene proceeds readily to the trialkylborane stage.   Oxidation yields a norborneol which by gas chromatographic examination is at least 99 per cent of the *exo* isomer.

This experiment suggests that hydroboration proceeds preferentially from the less hindered side of cyclic and bicyclic olefins with rigid structures.   This proposed generalization was tested by examining the hydroboration of α- and β-pinene.

The hydroboration of α-pinene proceeds to the dialkylborane stage. Oxidation of the organoborane provides pure isopinocampheol in excellent yield. In the same way, β-pinene is converted into *cis*-myrtanol. In both cases the products correspond to those expected on the basis of the proposed generalization.

Similarly, Wechter's conversion of cholesterol into cholestane-3β,-6α-diol[2] and the conversion of camphene into endo-camphenol by Dulou and Chrétien-Bessière[4] are also in accord with the proposed generalization.

Originally, these authors reported that they had obtained *trans*-myrtanol from the hydroboration-oxidation of β-pinene, in apparent contradiction with the generalization and in conflict with our own result. However, this discrepancy has now been resolved.[5] As has been pointed out, the organoboranes are readily isomerized at moderate temperatures under hydroboration conditions.[6,7] The organoborane initially produced in the hydroboration of β-pinene is oxidized to *cis*-myrtanol, in accordance with

[4] R. Dulou and Y. Chrétien-Bessière, *Bull. soc. chim. France*, **1959**, 1362.

[5] J. C. Braun and G. S. Fisher, *Tetrahedron Letters*, No. **21**, 9 (1960).

[6] H. C. Brown and B. C. Subba Rao, *J. Org. Chem.*, **22**, 1136 (1957).

[7] H. C. Brown and G. Zweifel, *J. Am. Chem. Soc.*, **82**, 1504 (1960).

the generalization. However, with moderate heating, the initial organo-borane is converted into an isomeric derivative which yields the less sterically hindered *trans*-myrtanol on oxidation.

The observation that the addition of the boron-hydrogen bond to β-pinene produces the less stable of the two possible isomers provides further evidence that the stereochemistry of the addition is controlled by the mechanism of the addition process and not by the stability of the addition product.

In further exploring the stereochemistry of the hydroboration reaction, we examined the results of treating norbornene with B_2H_6 and B_2D_6, treating the products with propionic acid, $CH_3CH_2CO_2H$, and the deuterated acid, $CH_3CH_2CO_2D$. This procedure yielded norbornane and its mono- and dideutero derivatives. Only two of the three groups undergo protonolysis under these conditions, but this causes no difficulty for the objectives of the experiment.[8]

$$\xrightarrow{\text{HB}} (C_7H_{11})_3B \xrightarrow{C_2H_5CO_2H} 2C_7H_{12} \qquad (I)$$

$$\xrightarrow{\text{HB}} (C_7H_{11})_3B \xrightarrow{C_2H_5CO_2D} 2C_7H_{11}D \qquad (II)$$

$$\xrightarrow{\text{DB}} (C_7H_{10}D)_3B \xrightarrow{C_2H_5CO_2H} 2C_7H_{11}D \qquad (III)$$

$$\xrightarrow{\text{DB}} (C_7H_{10}D)_3B \xrightarrow{C_2H_5CO_2D} 2C_7H_{10}D_2 \qquad (IV)$$

[8] H. C. Brown and K. J. Murray, *J. Org. Chem.*, **26**, 631 (1961).

The products were subjected to examination by NMR spectroscopy. The spectrum of norbornane (I) indicated in order of increasing field strength a peak of relative intensity 2.0, ascribed to the two tertiary bridgehead hydrogen atoms, a doublet (resulting from splitting by the axial hydrogens), ascribed to the four equatorial (*exo*) hydrogens with a relative intensity of 4.0, followed closely by an irregular peak of intensity 6.4, attributed to overlap of the two bridge hydrogens with the doublet from the four axial hydrogens.

The NMR spectra of the two monodeuterated norbornanes (II and III) were identical, with the relative intensity of the equatorial doublet decreased from 4.0 to 3.0. Finally, in the case of the dideuterated derivative (IV), the relative intensity of the equatorial hydrogen doublet decreased to 2.0, the relative intensities of the three peaks changing from the 2.0:4.0:6.4 value for norbornane (I) to 2.0:2.0:6.2 for the dideuteronorbornane (IV).

These results establish that the hydroboration-protonolysis reaction occurs in a *cis* manner involving the less hindered side of the double bond.

Again, without knowledge of the stereochemistry of the protonolysis reaction (Chapter 3), it is not possible to reach a definite conclusion as to the stereochemistry of the hydroboration process.

Stereochemistry of the Hydroboration Reaction

As has been pointed out, the mere fact that the over-all hydration and deuteration reactions correspond to a clean *cis* hydration or deuteration of the double bond does not permit one to draw conclusions as to the stereochemistry of each stage in the reaction. The observed stereochemistry is consistent either with a *cis* addition of the boron-hydrogen bond, followed by retention of configuration in the oxidation or protonolysis stages, or with a *trans* addition in the hydroboration step, with the oxidation or protonolysis proceeding with inversion of configuration.

Although the argument is not entirely rigorous, the results realized with norbornene strongly favor the first alternative — *cis* addition and oxidation or protonolysis with retention.

Consider the possibility that the hydration stage involves a *trans* addition of the boron-hydrogen bond to norbornene. It is generally accepted that the *exo* position is less hindered than the *endo*.[9] In that event, we should anticipate that the boron atom would prefer the less hindered *exo* position. Oxidation with inversion of configuration would then produce *endo*-norborneol, contrary to fact.

Similarly, the *trans* addition of the boron-deuterium bond to place the boron in the less hindered *exo* position would place the deuterium atom in the *endo* position, again contrary to the NMR spectrographic results.

[9] S. Beckmann and R. Mezger, *Ber.*, **89**, 2738 (1956).

On this basis it appears safe to conclude that the hydroboration reaction involves a simple *cis* addition of the boron-hydrogen bond to the less hindered side of the double-bond, followed by retention of configuration in both the oxidation and the protonolysis reaction.

It should also be mentioned here that the mechanisms suggested for the oxidation of organoboranes with alkaline hydrogen peroxide (Chapter 3) and for the protonolysis of organoboranes with carboxylic acids are consistent with the proposed retention of configuration in these stages. Moreover, there is growing evidence that retention of configuration constitutes the preferred reaction path for electrophilic substitutions on saturated carbon.[10,11]

[10] F. R. Jensen and L. H. Gale, *J. Am. Chem. Soc.*, **82**, 148 (1960).

[11] F. R. Jensen, L. D. Whipple, D. K. Wedegaertner, and J. A. Landgrebe, *J. Am. Chem. Soc.*, **82**, 2466 (1960).

The arguments presented have relied heavily on data obtained with the norbornene molecule. However, the stereochemical results realized in the hydroboration-oxidation of norbornene do not differ in any significant manner from those realized in numerous other systems. Consequently, there appears to be no reason to question the generality of the conclusions.

Applications

In the short time since the original announcement[1] numerous authors have applied this anti-Markownikoff *cis* hydration to a number of interesting systems, with gratifying, consistent results.

Thus, Winstein and his co-workers applied the procedure to an acyclic system, *cis-* and *trans*-2-*p*-anisyl-2-butene, and achieved a convenient synthesis of the diastereoisomeric alcohols, *threo-* and *erythro*-3-*p*-anisyl-2-butanol.[12]

Sondheimer and Wolfe applied the reaction to some decalin derivatives.[13] In the case of 10-methyl-$\Delta^{1(9)}$-octalin, two products were obtained, corresponding to hydroboration from both sides of the ring. However, 7,7,10-trimethyl-$\Delta^{1(9)}$-octalin yielded only one product. Evidently the *gem*-dimethyl group directs the hydroboration to the less hindered side of the ring system.

[12] E. L. Allred, J. Sonnenberg, and S. Winstein, *J. Org. Chem.*, **25**, 26 (1960).
[13] F. Sondheimer and S. Wolfe, *Canadian J. Chem.*, **37**, 1870 (1959).

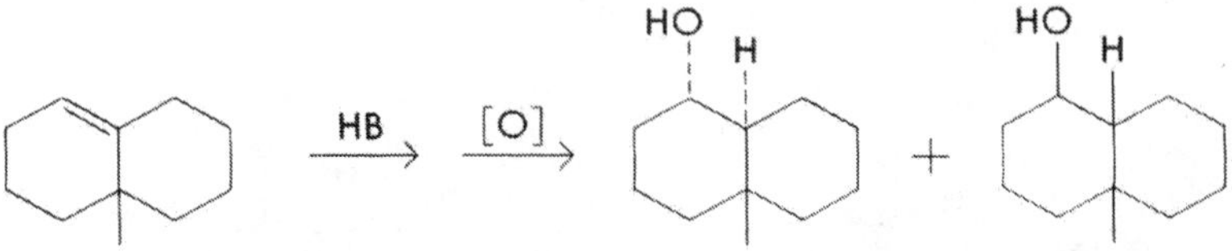

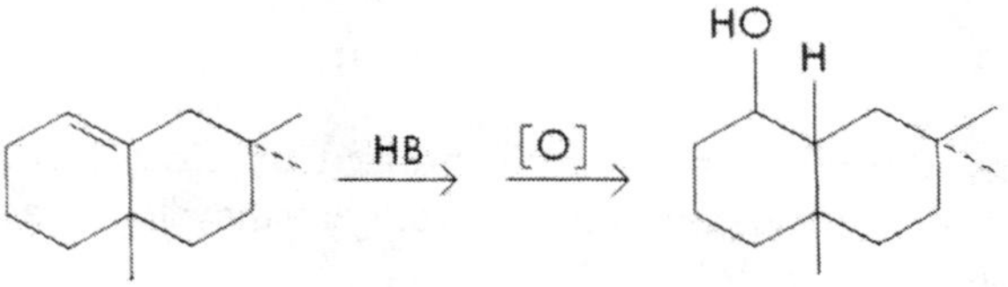

The rigid structure and well-established stereochemistry of the steroid system provides a particularly valuable testing ground for the proposed generalization. Wechter's conversion of cholesterol to cholestane-3β,-6α-diol[2] represents a *cis* hydration from the less hindered side. He also identified a minor constituent, coprostane-3β:6β-diol, which corresponds to a small quantity of *cis* hydration from the more hindered direction.

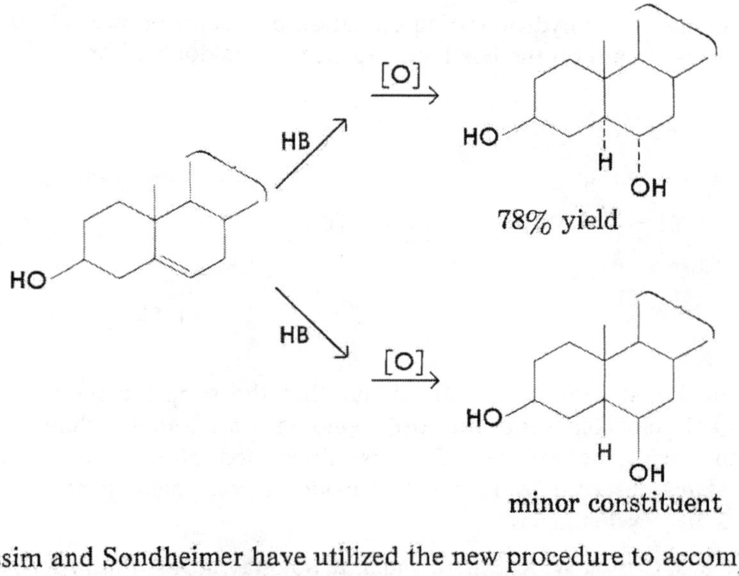

78% yield

minor constituent

Nussim and Sondheimer have utilized the new procedure to accomplish the synthesis of a number of stereochemically defined 11-oxygenated steroids.[14] For example, androst-9(11)-ene provided over 90 per cent of androstan-11α-ol.

[14] M. Nussim and F. Sondheimer, *Chem. & Ind. (London)*, **1960**, 400.

In bicyclic systems, the reaction has been utilized by Cristol and his co-workers to achieve the conversion of *exo*-1,2-dihydrodicyclopentadiene into the corresponding *exo* alcohol.[15]

Similarly, the hydroboration-oxidation of isodrin proceeds to place the hydroxyl group on the less hindered side of the double bond.[16,17]

Finally, it should be pointed out that the simple conversion of the 1-alkylcycloalkenes into the corresponding *trans*-2-alkylcycloalkanols, together with the preferential *trans* elimination of the 2-alkylcycloalkyl tosylates observed by Hückel,[18] provides a convenient general synthesis of 3-alkylcycloalkenes.[3]

[15] S. J. Cristol, W. K. Seifert, and S. B. Soloway, *J. Am. Chem. Soc.*, **82**, 2351 (1960).
[16] R. C. Cookson and E. Crundwell, *Chem. & Ind. (London)*, **1959**, 703.
[17] P. Bruck, D. Thompson, and S. Winstein, *Chem. & Ind. (London)*, **1960**, 405.
[18] W. Hückel and A. Hubele, *Ann.*, **613**, 27 (1958).

In conclusion, the simple *cis* addition of the boron-hydrogen bond to cyclic and bicyclic olefins provides an important new route to sterically defined derivatives. It may be anticipated that this reaction will play an increasingly important role in the synthesis of such derivatives.

9 | Isomerization

One of the more interesting and promising aspects of organoborane chemistry is the ready isomerization of organoboranes at moderate temperatures. In such isomerizations the boron atom exhibits the capacity of moving freely about the carbon structures of the alkyl or alicyclic groups of the organoborane, ending up at the least hindered position of the organic group. This isomerization occurs without any observable rearrangement or change in the basic carbon structure of the group.

$$
\begin{array}{ccccccc}
CH_3 & & CH_3 & & CH_3 & & CH_2{-}B{<} \\
| & & | & & | & & | \\
CH_2 & & CH_2 & & CH{-}B{<} & & CH_2 \\
| & & | & & | & & | \\
CH & \xrightarrow{HB} & CH{-}B{<} & \xrightarrow{\Delta} & CH_2 & \xrightarrow{\Delta} & CH_2 \\
\| & & | & & | & & | \\
CH & & CH_2 & & CH_2 & & CH_2 \\
| & & | & & | & & | \\
CH_2 & & CH_2 & & CH_2 & & CH_2 \\
| & & | & & | & & | \\
CH_3 & & CH_3 & & CH_3 & & CH_3
\end{array}
$$

It was observed by G. F. Hennion and his co-workers that secondary organoboranes, synthesized *via* the usual Grignard procedure, undergo a slow isomerization into the primary derivatives at reflux temperatures. Thus they report that tri-*sec*-butylborane is transformed into tri-*n*-butylborane under reflux at 200 to 215° for 48 hours.[1]

[1] G. F. Hennion, P. A. McCusker, E. C. Ashby, and A. J. Rutkowski, *J. Am. Chem. Soc.*, **79**, 5194 (1957).

$$
\underset{\substack{|\\\text{CH}_3}}{\text{CH}_3\text{CH}_2\text{CH}}\text{—)}_3\text{B} \xrightarrow[\substack{48 \text{ hr}}]{200\text{–}215°} \underset{\substack{|\\\text{CH}_3\text{CH}_2\text{CH}_2}}{\text{CH}_2}\text{—)}_3\text{B}
$$

In our early hydroboration experiments, we encountered phenomena which suggested that isomerization of the organoboranes was far more facile under hydroboration conditions.[2,3] For example, in our early experiments we isolated the triorganoboranes by distillation from the diglyme solution utilized for the hydroboration. The isolated product was then oxidized with alkaline hydrogen peroxide to obtain the alcohol. In this way we observed that the hydroboration of 2-pentene yields a tri-*sec*-pentylborane, b.p. 81 to 82° at 2 mm, which oxidized to 63 per cent 2-pentanol and 37 per cent 3-pentanol.

In later experiments we developed a simplified procedure in which the organoborane was oxidized in the diglyme solution, *in situ*, without isolation. In this procedure, the organoborane from 2-pentene yielded 2- and 3-pentanol in essentially equimolar amounts (Chapter 7). It was evident that during the earlier distillation a partial shift of the boron atom from the 3 to the 2 position of the alkyl group had occurred, similar to that described by Hennion and his co-workers. However, our product had been subjected to relatively mild temperatures (~ 100°) for relatively short periods of time (~ 2 hours).

Study of this phenomenon has revealed that the isomerization reaction is strongly catalyzed by small amounts of diborane or other moieties containing boron-hydrogen bonds.[4] The recommended hydroboration procedure utilizes 10 to 20 per cent excess "hydride" to ensure the quantitative utilization of the olefin. It is this excess hydride which appears to be largely responsible for the enhanced rate.

For example, the hydroboration of 2-hexene proceeds to place approximately 50 per cent of the boron on each of the 2 and 3 positions. Oxidation of the product yields equimolar amounts of 2- and 3-hexanol. (In using formulae to show the isomerization it is convenient to show the carbon skeletons and to omit the hydrogen. As already mentioned, the boron atom appears to move freely about the basic carbon structure without altering it.) However, if the diglyme solution is heated at 150° for 1 hour, oxidation of the contained organoborane gives 88 per cent 1-, 7 per cent 2-, and 5 per cent 3-hexanol.

[2] H. C. Brown and B. C. Subba Rao, *J. Org. Chem.*, **22**, 1136 (1957).

[3] H. C. Brown and B. C. Subba Rao, *J. Am. Chem. Soc.*, **81**, 6434 (1959).

[4] H. C. Brown and G. Zweifel, manuscript in preparation.

C—C—C—C=C—C

HB $\downarrow$ 10–20% excess H$^-$

$$C—C—C—\underbrace{C—C}—C \quad \xrightarrow[\text{1 hr}]{150°} \quad C—C—C—C—C—C$$

B $\bigwedge$ B $\bigwedge$

$\downarrow$ [O] $\downarrow$ [O]

50%   C—C—C—C—C—C   88%   C—C—C—C—C—C

OH   OH

+

50%   C—C—C—C—C—C

OH

However, if a slight excess of 2-hexene is utilized, the isomerization is far slower under otherwise identical conditions: 13 per cent 1-, 64 per cent 2-, 23 per cent 3-hexanol.

This observation that the rate of isomerization can be controlled in this way has very important practical implications. It becomes possible to isomerize an organoborane in the presence of minute quantities of excess "hydride" and then to add an olefin to achieve a displacement reaction (Chapter 10), without concern that any back isomerization will occur at the temperature required for the displacement.

The usual procedure is a very simple one. A mixture of the olefin and sodium borohydride (10 per cent excess) in diglyme is treated with boron trifluoride to achieve hydroboration. The reaction mixture is heated to reflux (diglyme, b.p. 162°) and maintained there for 1 hour. Isomerization is essentially complete in all cases except where the aliphatic chain is relatively long.

The isomerizations observed by Hennion and his co-workers involve a shift of the boron atom from the secondary or tertiary 2 position of a carbon chain to the adjacent primary position.[1] Our own early experiments with the organoboranes from 2-pentene and 2-hexene had indicated that the boron atom was capable of moving in steps from the 3 to the 2 and finally to the 1 position of the chain. It was of interest to learn whether this migration could be achieved down a relatively long chain, whether the boron atom could pass alkyl branches in the chain, and whether it could be moved out of an alicyclic ring into the side chain.

Accordingly, a detailed study was undertaken of the isomerization of the organoboranes from a large number of olefins in order to establish the scope and possible limitations.[4,5]

Organoboranes from Straight-Chain Olefins

The hydroboration of 3-hexene yields an organoborane which contains all the boron attached at the 3 position. After 1 hour at 150°, 90 per cent of the boron is found in the 1 position, with 6 per cent in the 2 and 4 per cent in the 3 position.[4]

In order to slow down the reaction so that observations could be made on intermediate stages, the reaction temperature was dropped to 125°. The results clearly reveal a stepwise shift from the 3 to the 2 to the 1 position (Table 9-1).

Table 9-1

Isomerization of Organoborane from 3-Hexene at 125°

Time, hr.	Yield, %		
0	100	0	0
1	26	30	44
2	18	25	57
4	11	15	74
8	9	9	82
24	6	6	88

It is of interest that the related isomerization of the organoborane from 1-hexene reveals a growth of small quantities of the 3-hexyl derivative. Consequently, we appear to realize an equilibrium distribution of the boron at all positions of the chain, but with a very strong preference for the terminal position.

[5] H. C. Brown and G. Zweifel, *J. Am. Chem. Soc.*, **82**, 1504 (1960).

Finally, it proved possible to take a mixture of 2-, 3-, 4-, 5-decenes and to convert it to an 80 per cent yield of 1-decanol via hydroboration-isomerization-oxidation. Similarly, a mixture of tetradecenes (excluding the 1-isomer) was converted to a 70 per cent yield of 1-tetradecanol.[2,3] These more extensive isomerizations required longer isomerization periods — 4 hours for the decenes and 18 hours for the tetradecenes. However, these experiments were performed before we had recognized the catalytic influence of excess boron-hydride moieties; such long reaction periods may not be essential under conditions of controlled catalysis.

Organoboranes from Branched-Chain Olefins

The hydroboration of 2,4,4-trimethyl-3-pentene proceeds to place the boron atom at the 3 position. The product isomerizes with remarkable ease to place the boron atom at the terminal position. Consequently, the boron atom experiences no difficulty in moving past a methyl branch.[5]

Indeed, in attempting to measure the rate, it proved necessary to lower the temperature down to 75°; even at this temperature the reaction was quite fast (Table 9–2).[4]

It should be recalled that the hydroboration of this olefin proceeds to the dialkylborane stage. Consequently, there is present a large excess of the catalytic boron-hydrogen species. This is presumably responsible in part for the fast rate.

The fact that the boron atom migrates easily past a methyl branch is also indicated by the following transformation.

Under these conditions, the boron atom does not migrate past a double branch. For example, in the product from 4-methyl-2-pentene, the boron atom moves to both ends of the chain, giving 59 per cent of 4-methyl-1-

Table 9-2

Isomerization of Organoborane from 2,4,4-Trimethyl-2-pentene at 75°

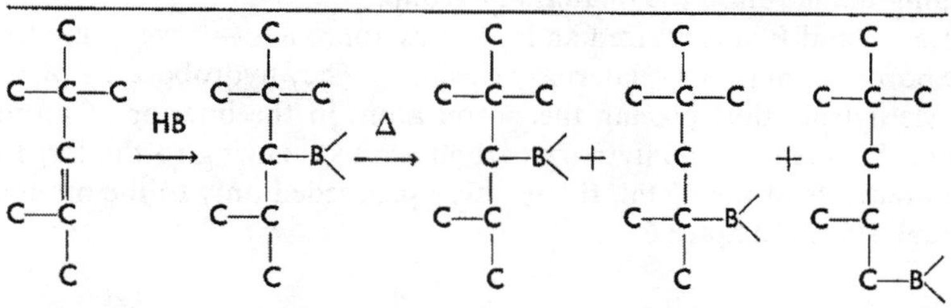

Time, hr.		Yield, %	
0	98	2	0
2	5	20	75
4	5	9	86
8	1	3	96
24	Trace	2	98

pentylborane and 39 per cent of 2-methyl-1-pentylborane. However, in the isomerization of the related organoborane from 4,4-dimethyl-2-pentene, the boron atom moves only to the unsubstituted end of the chain.

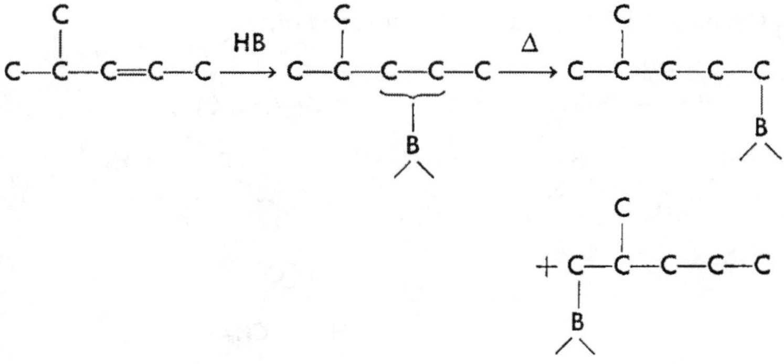

59%

39%

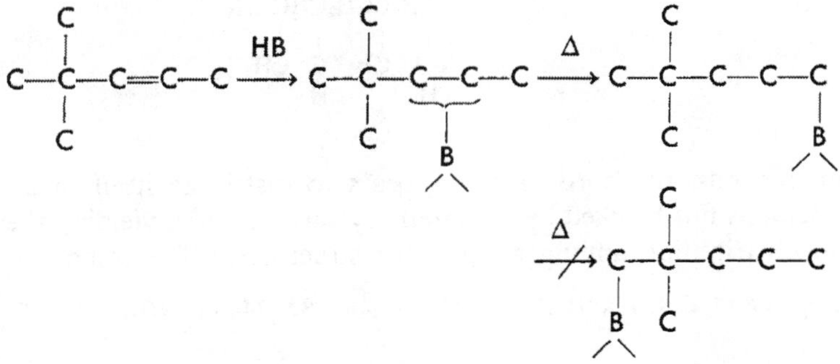

No evidence is observed for the formation of the isomeric derivative containing the boron in the quaternary group.

Logan and Flautt utilized an ingenious approach — attempting to force the boron atom past a quaternary carbon.[6] They hydroborated *trans*-di-*t*-butylethylene, thus placing the boron atom in the interior of an alkane chain, between *two* quaternary carbon atoms. Owing to the large steric requirements of the olefin, the reaction proceeded only to the monoalkylborane stage (Chapter 6).

However, when this monoalkylborane was heated, it lost hydrogen under relatively mild conditions and formed a boron heterocycle. Oxidation yielded a glycol, 2,2,5,5-tetramethyl-1,4-hexanediol.

In isomerizations, the boron atom appears to distribute itself in all possible positions not blocked by a quaternary carbon atom, yielding the equilibrium distribution among all possible structures. The same dis-

[6] T. J. Logan and T. J. Flautt, *J. Am. Chem. Soc.*, **82**, 3446 (1960).

tribution is realized, starting from each of the possible isomers. The data for a typical system are summarized in Table 9-3.

Table 9-3

Isomerization of the Organoboranes from 2-Methyl-1-butene, 2-Methyl-2-butene, and 3-Methyl-1-butene

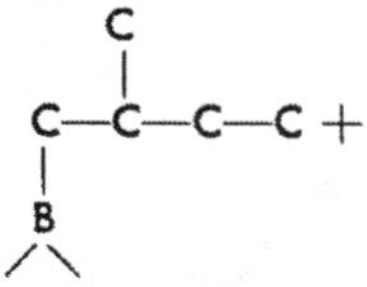

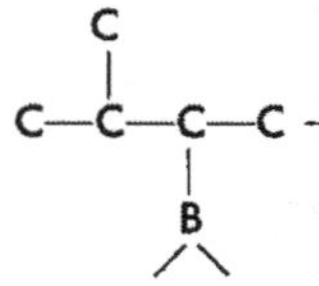

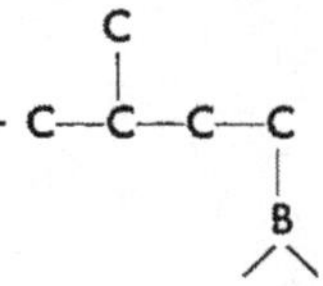

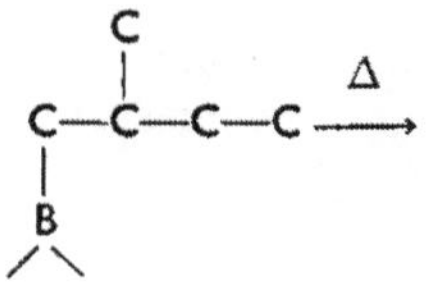 40%	1%	59%
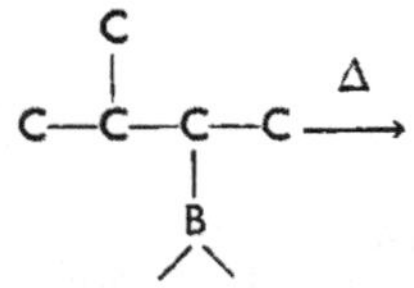 40	2	58
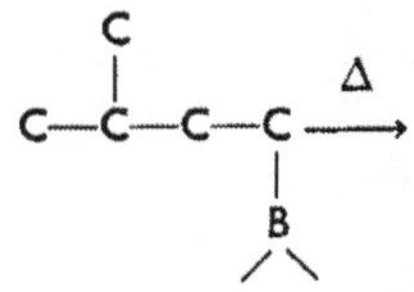 41	1	58

It is apparent that in these isomerizations the boron atom prefers the terminal position. In cases in which the two terminal positions are not identical, as in the 2-methylbutyl and the 2-methylpentyl systems, the boron atom prefers the less hindered position.

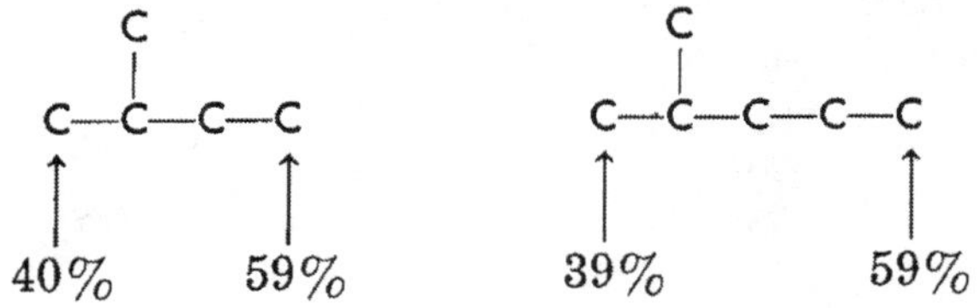

Consequently, it appears probable that steric influences in the trialkyl-borane derivative direct the boron atom to the position that minimizes the steric interactions.

Organoboranes from Cyclic Olefins

The isomerization of organoboranes from cyclic olefins appears to offer no major points of difference from the cases just discussed.[7] The boron atom moves with ease around the ring and into the side chain. For example, the same equilibrium mixture is obtained in the isomerization of the organoborane from 1-methylcyclohexene and 4-methylcyclohexene.

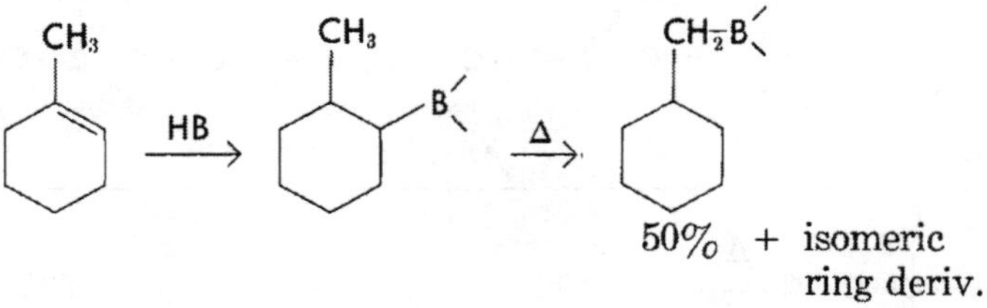

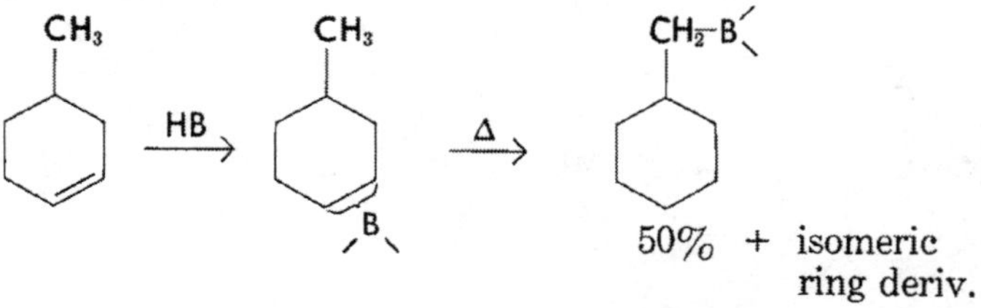

50% + isomeric
      ring deriv.

50% + isomeric
      ring deriv.

With a longer side chain, there is a greater tendency for the boron atom to accumulate in the side chain.

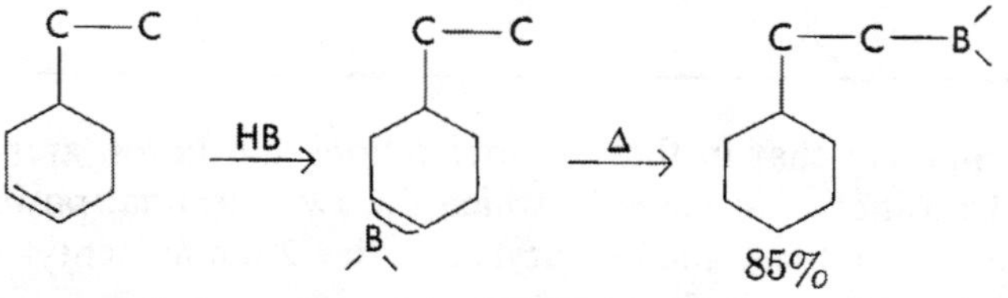

85%

Even the organoboranes from bicyclic olefins, such as α-pinene, undergo facile isomerization to place the boron in the exocyclic position. Oxidation produces *trans*-myrtanol.

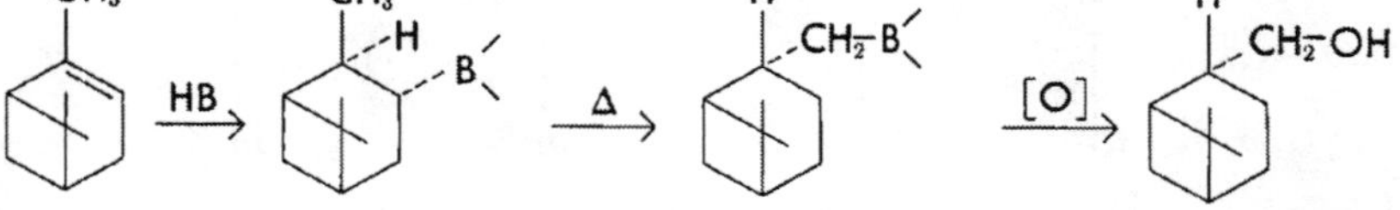

---

[7] H. C. Brown, M. V. Bhatt, and G. Zweifel, manuscript in preparation.

On the other hand, oxidation of the organoborane obtained in the hydroboration of β-pinene at 0° produces an isomeric derivative, *cis*-myrtanol.

It has recently been demonstrated that at moderate temperatures there is a
facile isomerization from the *cis*-myrtanylborane structure to the more
stable *trans*-myrtanylborane system.[8]

Organoboranes from Dienes

The isomerization of organoboranes from dienes offers a number of
points of unusual interest.[9]   However, discussion will be deferred to the
chapter dealing with the hydroboration of dienes (Chapter 15).

Mechanism

The addition of the boron-hydrogen bond to olefins appears to involve
a simple 4-center *cis* addition.[10,11]

---

[8] J. C. Braun and G. S. Fisher, *Tetrahedron Letters*, No. **21**, 9 (1960).
[9] K. A. Saegebarth, *J. Am. Chem. Soc.*, **82**, 2081 (1960).
[10] H. C. Brown and G. Zweifel, *J. Am. Chem. Soc.*, **82**, 4708 (1960).
[11] H. C. Brown and G. Zweifel, *J. Am. Chem. Soc.*, **83**, 2544 (1961).

From Rosenblum's observation, it is evident that this addition must be partially reversible at elevated temperatures.[12,13]

$$\underset{\underset{B-}{\overset{|}{\underset{|}{H}}}{\overset{|}{\underset{}{C}}}-\underset{\underset{|}{\overset{|}{B-}}}{\overset{|}{C}} \rightleftharpoons \underset{\underset{|}{H-B-}}{\overset{|}{C}}{=}\underset{}{\overset{|}{C}}$$

Originally, we postulated that the isomerization of the boron atom from an internal position on a carbon chain to the terminal position proceeded through a succession of such eliminations and additions.

This mechanism accounts for the observed ease in moving the boron atom past a single alkyl branch and the difficulty in moving it through a double branch. It accounts also for the effect of excess olefin in repressing the isomerization reaction. However, in its present form it fails to account for the catalytic effect of boron-hydrogen bonds.

Recent studies have shown that the dialkylboranes exist as dimers in ether solvents[14] (Chapter 12).

[12] L. Rosenblum, *J. Am. Chem. Soc.*, **77**, 5016 (1955).
[13] R. Köster, *Ann.*, **618**, 31 (1958).
[14] H. C. Brown and G. J. Klender, *J. Inorg. Chem.*, in press.

A study of the reaction of disiamylborane, actually *sym*-tetrasiamyldiborane, with representative olefins has revealed that the reaction is second order, first order in *sym*-tetrasiamyldiborane and first order in olefin[15] (Chapter 13).

The following transition state has been suggested for the reaction.

Presumably the solvent is involved in solvating the leaving group, R_2BH.

If the addition reaction involves the dimeric diborane, the reverse reaction, elimination of R_2BH, must involve the same intermediate. Consequently, it is proposed that the isomerization proceeds through successive eliminations and additions, as originally suggested, but with the modification that it is a dimeric diborane species which participates in both the additions and eliminations.

With this modification it is possible to account both for the inhibition by olefin and the catalysis by boron-hydrogen moieties.

Cyclization

It was pointed out that the hydroboration of 2,4,4-trimethyl-2-pentene proceeds to the formation of the dialkylborane as a limit. This product isomerizes very rapidly, even at temperatures as low as 75° (see Table 9-2). At these temperatures we realized nearly quantitative yields of 2,4,4-trimethyl-1-pentanol after oxidation. However, when the isomerization was carried out at the usual conditions (162°), hydrogen was evolved and the yield dropped.

Examination of the reaction mixture revealed the presence of a glycol, 2,2,4-trimethyl-1,5-pentanediol.[16] The reaction is presumably related to the cyclizations observed by Winternitz and Carotti,[17] Logan and Flautt,[6] and Köster and Rotermund[18] (Chapter 3).

[15] H. C. Brown and A. W. Moerikofer, *J. Am. Chem. Soc.*, **83**, 3417 (1961).
[16] H. C. Brown, K. J. Murray, and G. Zweifel, manuscript in preparation.
[17] P. F. Winternitz and A. A. Carotti, *J. Am. Chem. Soc.*, **82**, 2430 (1960).
[18] R. Köster and G. Rotermund, *Angew. Chem.*, **72**, 138 (1960).

This internal substitution reaction appears to be especially facile with a tertiary butyl group in position to form a five- or six-membered boron heterocycle.[6] It appears that primary carbon-hydrogen bonds are especially favorable for the reaction,[17,18] and the high effectiveness of the tertiary butyl group may derive primarily from the large number of primary carbon-hydrogen bonds that this group provides in suitable reaction position.   The cyclization reaction is much less significant in isomerizations involving other types of structures, such as the normal alkanes. Although conditions can be utilized to bring these groups to participate in a cyclization reaction,[16,18] no special precautions need to be observed with these groups to avoid this reaction during the usual isomerization procedure.

In the case of 2,4,4-trimethyl-2-pentene (or 2,4,4-trimethyl-1-pentene) it is possible, by working at higher temperatures (160 to 200°), to realize excellent yields of the glycol (80 per cent, based on the equation indicated). By working at lower temperatures, or by utilizing nearly the theoretical stoichiometric ratio (3 molecules of 2,4,4-trimethyl-2-pentene per BH_3), it is possible to avoid this reaction and to achieve nearly quantitative yields of the 1 derivative.

Conclusion

The isomerization reaction significantly extends the utility of the hydroboration reaction.   The structure of the organoborane obtained is no longer limited by the position of the double bond in the olefin.   Consequently, it becomes possible to synthesize many derivatives not immediately related to the structure of the initial olefin.

As will be pointed out in Chapter 10, it is possible to convert organoboranes into the corresponding olefins by treating with another olefin — the displacement reaction.   One of the fascinating possibilities opened up by a combination of hydroboration-isomerization-displacement is the feasibility of a simple contra-thermodynamic isomerization of olefins.

10 | Displacement Reactions

The rapid and facile isomerization of the boron atom from an internal position on a carbon chain to the terminal position was originally attributed to a succession of eliminations and additions at the reaction temperature[1,2] (Chapter 9).

$$
\begin{array}{cc}
\begin{array}{c}
\text{H} \quad \text{H} \quad \text{H} \\
\text{R}-\overset{|}{\underset{|}{\text{C}}}-\overset{|}{\underset{|}{\text{C}}}-\overset{|}{\underset{}{\text{C}}}-\text{H} \\
\text{B} \quad \text{H} \quad \text{H} \\
/\backslash
\end{array}
&
\begin{array}{c}
\text{H} \quad \text{H} \quad \text{H} \\
\text{R}-\overset{|}{\underset{|}{\text{C}}}-\overset{|}{\underset{|}{\text{C}}}-\overset{|}{\underset{}{\text{C}}}-\text{H} \\
\text{H} \quad \text{H} \quad \text{B} \\
/\backslash
\end{array}
\end{array}
$$

$$\Updownarrow \qquad\qquad\qquad\qquad \Updownarrow$$

$$
\begin{array}{c}
\text{H} \quad \text{H} \quad \text{H} \\
\text{R}-\text{C}=\text{C}-\overset{|}{\underset{|}{\text{C}}}-\text{H} \\
\text{B}-\text{H} \quad \text{H} \\
/\backslash
\end{array}
\rightleftharpoons
\begin{array}{c}
\text{H} \quad \text{H} \quad \text{H} \\
\text{R}-\overset{|}{\underset{|}{\text{C}}}-\overset{|}{\underset{|}{\text{C}}}-\overset{|}{\underset{}{\text{C}}}-\text{H} \\
\text{H} \quad \text{B} \quad \text{H} \\
/\backslash
\end{array}
\rightleftharpoons
\begin{array}{c}
\text{H} \quad \text{H} \quad \text{H} \\
\text{R}-\overset{|}{\underset{}{\text{C}}}-\text{C}=\text{C}-\text{H} \\
\text{H} \quad \text{B}-\text{H} \\
/\backslash
\end{array}
$$

It is evident from this mechanism that the presence of another olefin in the reaction mixture of similar or greater reactivity should cause it to participate, freeing a corresponding number of molecules of the original olefin used to form the organoborane.

[1] H. C. Brown and B. C. Subba Rao, *J. Org. Chem.*, **22**, 1136 (1957).
[2] H. C. Brown and B. C. Subba Rao, *J. Am. Chem. Soc.*, **81**, 6434 (1959).

$$R'CH{=}CH_2 + RCH{-}CHCH_3 \overset{\Delta}{\rightleftharpoons} R'CH{=}CH_2 + RCH{=}CHCH_3$$

Such a displacement reaction would lead to an equilibrium mixture of all possible organoboranes. It is obvious that such an equilibrium could be driven toward completion (1) by using a large concentration of the displacing olefin, (2) by distilling the more volatile olefin out of the reaction mixture, and (3) by using an olefin that forms a very stable organoborane. All these methods have been applied.

In the laboratory it is convenient to use a less volatile olefin to displace a more volatile olefin. For example, in our first exploration of this reaction, tri-*n*-pentylborane was heated with 1-decene in a flask attached to a distilling column. A good yield of 1-pentene was realized and the product was identified as tri-*n*-decylborane.[1,2]

$$3n\text{-}C_8H_{17}CH{=}CH_2 + (n\text{-}C_5H_{11})_3B \xrightarrow{160°} 3n\text{-}C_8H_7CH{=}CH_2\uparrow + (n\text{-}C_{10}H_{21})_3B$$

A similar displacement reaction in the organoaluminum derivatives was discovered some time ago by Ziegler and his co-workers[3] and has been extensively explored.[4] Perhaps the most noteworthy difference is the fact that the organoaluminum compounds undergo chain-growth reactions under conditions very similar to those required for the displacement. Consequently, very careful control of the experimental conditions are required to achieve reasonable yields in the displacement.[5]

The organoboranes do not undergo chain growth under displacement conditions. Consequently, it is readily possible to achieve essentially quantitative conversions without special precautions.

[3] K. Ziegler, *Angew. Chem.*, **64**, 323 (1952).

[4] K. Ziegler, W. R. Kroll, W. Larbig, and O. W. Stendel, *Ann.*, **629**, 53 (1960).

[5] K. Ziegler, Organo-Aluminum Compounds, in H. Zeiss (ed.), "Organometallic Chemistry," Chap. 5, American Chemical Society Monograph Series, Reinhold Publishing Corporation, New York, 1960.

The displacement reaction of organoboranes has been studied intensively in Ziegler's laboratory by Köster as a convenient synthesis of organoboranes.[6] Triisobutylborane, now readily available from the reaction of triisobutylaluminum and methyl borate, was heated with a number of olefins to displace isobutylene, forming the corresponding organoborane. In this way he has synthesized a number of organoboranes, such as tri-*n*-decylborane, tricyclohexylborane, and tri-(2-phenylpropyl)-borane.

$$3n\text{-}C_8H_{17}CH{=}CH_2 + [(CH_3)_2CHCH_2]_3B \longrightarrow$$
$$(n\text{-}C_8H_{17}CH_2CH_2)_3B + 3(CH_3)_2C{=}CH_2$$

$$3C_6H_{10} + [(CH_3)_2CHCH_2]_3B \longrightarrow (C_6H_{11})_3B + 3(CH_3)_2C{=}CH_2$$

$$3C_6H_5C(CH_3){=}CH_2 + [(CH_3)_2CHCH_2]_3B \longrightarrow$$
$$[C_6H_5CH(CH_3)CH_2]_3B + 3(CH_3)_2C{=}CH_2$$

The emphasis has been quite different in our investigations.[7-9] The hydroboration procedures we have developed for laboratory applications are highly convenient in that they proceed from stable, easily handled chemicals[10] (Chapter 5). Consequently, our interest in the displacement reaction has been primarily in its utility as a new type of elimination reaction which might be combined with isomerization to provide a new, convenient synthesis of olefins. For example, a simple conversion of the following type would be quite useful.

[6] R. Köster, *Ann.*, **618**, 31 (1958).

[7] H. C. Brown and M. V. Bhatt, *J. Am. Chem. Soc.*, **82**, 2074 (1960).

[8] H. C. Brown and M. V. Bhatt, manuscript in preparation.

[9] H. C. Brown, M. V. Bhatt, and G. Zweifel, manuscript in preparation.

[10] H. C. Brown, K. J. Murray, L. J. Murray, J. A. Snover, and G. Zweifel, *J. Am. Chem. Soc.*, **82**, 4233 (1960).

The displacement reaction is carried out in the same temperature range as isomerization.   In order that the displacement reaction should be useful for the synthesis of olefins, it was necessary to establish that back-isomerization of the organoborane does not occur during the displacement.   It was soon observed that the ease of displacement varies considerably with olefins of different structural types.   Finally, it was of interest to learn the direction taken by the elimination in organoboranes capable of losing a boron-hydrogen bond in two directions.

These considerations made it desirable to undertake a systematic survey of the displacement of olefins from organoboranes synthesized by the hydroboration of representative structural types.   The procedure was simple.   The olefin in question was converted to the organoborane in diglyme solution by treatment with sodium borohydride–boron trifluoride. An excess of a less volatile olefin, usually 1-decene, was added.   The flask was attached to a fractionating column and the reaction mixture heated to reflux.   Olefin was collected.   The amount of olefin collected in intervals of time and the composition of the product (gas chromatography) were noted.   The yields averaged 80 ± 10 per cent.   Actually, the yields are probably nearly quantitative with the observed values arising primarily from the slowness of the reaction in the latter stages and the losses in the column.

Monosubstituted Terminal Olefins, $RCH{=}CH_2$

Terminal olefins, such as 1-pentene, 1-hexene, 3-methyl-1-butene, and 3-ethyl-1-pentene, require approximately 6 to 8 hours for essential completion of the displacement stage.   The products are 90 to 95 per cent pure terminal olefins, with 5 to 10 per cent of the 2-alkene.

It was noted that the 2- isomer largely accumulates in the initial fractions of the product, the latter fractions being almost pure 1-alkene.   Apparently, the minor product arises from the 6 to 7 per cent of the *sec*-alkylboron derivative formed in the hydroboration of a terminal olefin (Chapter 7), with this isomer undergoing displacement more rapidly than the *n*-alkylborane present.   It was established in independent experiments that pure *sec*-alkylboranes undergo displacement considerably more rapidly than the primary derivatives.

As will be discussed later, the ease of displacement varies considerably with the structure of the olefin and can be correlated with the relative stability of the double bond, as indicated by its heat of hydrogenation.

In a typical experiment the organoborane from the hydroboration of 1-pentene undergoes displacement with 1-decene. In 5 hours an 83 per cent yield of pentene is realized, with analysis indicating the presence of 91 per cent 1-pentene and 9 per cent 2-pentene.

$$CH_3CH_2CH_2CH{=}CH_2 \xrightarrow{HB} \underset{\underset{94\%}{\overset{/\backslash}{B}}}{CH_3CH_2CH_2CH_2CH_2} + \underset{\underset{6\%}{\overset{/\backslash}{B}}}{CH_3CH_2CH_2CHCH_3}$$

$$5\ hr \downarrow RCH{=}CH_2$$

$$\underset{91\%}{CH_3CH_2CH_2CH{=}CH_2} + \underset{9\%}{CH_3CH_2CH{=}CHCH_3}$$

The presence of an alkyl substituent in the 3 position, as in 3-methyl-1-butyl- and 3-ethyl-1-pentylborane, does not appear to have any significant influence on the reaction. In both cases the product consists of the terminal olefin predominantly, with a small quantity of the 2-isomer.

$$(C_2H_5)_2CHCH{=}CH_2 \xrightarrow{HB} \underset{\overset{/\backslash}{B}}{(C_2H_5)_2CHCH_2CH_2} + \underset{\overset{/\backslash}{B}}{(C_2H_5)_2CHCHCH_3}$$

$$\downarrow RCH{=}CH_2$$

$$\underset{90\%}{(C_2H_5)_2CHCH{=}CH_2} + \underset{10\%}{(C_2H_5)_2C{=}CHCH_3}$$

Disubstituted Terminal Olefins, $R_2C{=}CH_2$

The considerably greater ease of displacement of olefins of the isobutylene type had previously been observed in the corresponding reaction of the aluminum alkyls.[3] The displacement of 2-methyl-1-butene is complete in 3 hours, with the product consisting of 99 per cent of 2-methyl-1-butene and 1 per cent of 2-methyl-2-butene. Similarly, the displacement of 2,4,4-trimethyl-1-pentene proceeds readily, with the purity of the displaced olefin indicated as essentially 100 per cent of the terminal isomer.

$$CH_3CH_2\overset{\overset{\displaystyle CH_3}{|}}{C}\!\!=\!\!CH_2 \xrightarrow{\ HB\ } CH_3CH_2\overset{\overset{\displaystyle CH_3}{|}}{\underset{\underset{\displaystyle 99\%}{\underset{\displaystyle \wedge}{B}}}{C}}HCH_2 + CH_3CH_2\overset{\overset{\displaystyle CH_3}{|}}{\underset{\underset{\displaystyle 1\%}{\underset{\displaystyle \wedge}{B}}}{C}}CH_3$$

$$3\ hr \Big\downarrow RCH\!\!=\!\!CH_2$$

$$CH_3CH_2\overset{\overset{\displaystyle CH_3}{|}}{C}\!\!=\!\!CH_2 + CH_3CH\!\!=\!\!\overset{\overset{\displaystyle CH_3}{|}}{C}CH_3$$
$$99\% \qquad\qquad 1\%$$

Disubstituted Internal Olefins, RCH=CHR′

The displacement reaction involving 2-pentene is also relatively fast, an 87 per cent yield being realized in 2.5 hours. The hydroboration of 2-pentene proceeds to place approximately 50 per cent of the boron on each of the two carbon atoms of the double bond. Evidently the elimination prefers to proceed to give the more stable of the two possible olefins, since the product contains only 2 per cent of 1-pentene.

$$CH_2CH_2CH\!\!=\!\!CHCH_3 \xrightarrow{\ HB\ } CH_3CH_2CH_2\overset{}{\underset{\underset{\underset{\displaystyle 50\%}{\displaystyle \wedge}}{B}}{C}}HCH_3 + CH_3CH_2\overset{}{\underset{\underset{\underset{\displaystyle 50\%}{\displaystyle \wedge}}{B}}{C}}HCH_2CH_3$$

$$\Big\downarrow RCH\!\!=\!\!CH_2$$

$$CH_3CH_2CH\!\!=\!\!CHCH_3 + CH_3CH\!\!=\!\!CHCH_2CH_3$$
$$+\ CH_3CH_2CH_2CH\!\!=\!\!CH_2 \quad (2\%)$$

Trisubstituted Olefins, R₂C=CHR

Trisubstituted olefins, such as 2-methyl-2-butene, 3-ethyl-2-pentene, and 2,4,4-trimethyl-2-pentene, undergo hydroboration to place approximately 98 per cent of the boron on the secondary position. Displacement is quite rapid. In each case the product is predominantly the original olefin.

$$\underset{\underset{CH_3C=CHCH_3}{|}}{CH_3} \xrightarrow{\text{HB}} \underset{\underset{CH_3CHCHCH_3}{|}}{CH_3} + \underset{\underset{CH_3CCH_2CH_3}{|}}{CH_3}$$

$$\underset{98\%}{\overset{B}{\diagdown}} \qquad \underset{2\%}{\overset{B}{\diagdown}}$$

$$\Big\downarrow \text{RCH}{=}\text{CH}_2$$

$$\underset{\underset{CH_3C=CHCH_3}{|}}{CH_3} + \underset{\underset{CH_2=CCH_2CH_3}{|}}{CH_3}$$

$$99\% \qquad\qquad 1\%$$

Cyclic Olefins

Cyclopentene and cyclohexene undergo displacement in purities approaching 100 per cent.

It is particularly important that olefins such as methylenecyclohexane and β-pinene undergo ready displacement without any evidence of rearrangement to an isomeric structure.

Only in the case of the 1-methylcycloalkenes is there observed a major tendency for the elimination to take place in two directions.

77%   23%

Even though the conversion to the 3-alkylcycloalkenes is relatively low, only 20 per cent, the two isomers are easily separable by fractional distillation. The reaction, therefore, appears to provide a simple route from the readily available 1-alkylcycloalkenes to the less available 3-alkylcycloalkenes.

Bicyclic Olefins

α-Pinene undergoes displacement without unusual difficulty, the product consisting of 90 per cent α-pinene, 5 per cent β-pinene, and 5 per cent of a third component, presumably δ-pinene. The formation of the β-pinene is unexpected. It may indicate that at the temperature used, 160°, a small amount of isomerization of the organoborane may have occurred.

The results with norbornene were remarkable. There was no evidence of its displacement with 1-decene, even after 15 hours at 160°.

The Displacing Olefin

The above studies establish that the relative ease of displacement of olefins from the organoboranes is

$$R_2C=CHR > R_2C=CH_2, \quad RCH=CHR > RCH=CH_2$$

This order corresponds roughly to the thermodynamic stability of the olefins, as measured by their heats of hydrogenation.[11] Since the reaction appears to proceed through an equilibrium established between the two olefins in the system and the intermediate organoboranes, it is not surprising that the efficacy of the displacing olefin is largest for 1-octene, drops sharply for 2-octene and 2,2,4-trimethyl-1-pentene, and is very poor for 2,4,4-trimethyl-2-pentene.

[11] G. B. Kistiakowsky, H. Romeyn, Jr., J. R. Ruhoff, H. A. Smith, and W. E. Vaughan, *J. Am. Chem. Soc.*, **57**, 65 (1935).

In the previous section it was reported that there was no evidence for the displacement of norbornene by 1-decene. This indicates that norbornene should be a very powerful displacing olefin, which might be quite convenient for laboratory work to avoid the requirement of distillation to achieve high conversions. Ethylene should also be a very efficient displacing olefin. Its use in the laboratory suffers from the disadvantage of requiring pressure equipment.

Rough experiments also indicate that the rate of displacement is independent of the concentration of the displacing olefin. The time required to realize an 80 per cent yield of 1-pentene from tri-n-pentylborane was the same whether one or two equivalents of 1-decene were present in the refluxing reaction mixture.

Mechanism

All the results are in accord with the mechanism originally suggested. At the temperature of displacement there is a slow dissociation into dialkylborane and olefin. The presence of a large concentration of relatively active olefin in the reaction mixture combines with the intermediate dialkylborane, liberating the olefin originally contained in the trialkylborane.

This displacement reaction is considerably slower than the isomerization reactions described in Chapter 9. At first sight, this appears anomalous. According to the proposed interpretation, isomerization must require many successive eliminations and additions before the boron atom finally attains the end of the chain. On the other hand, with a large excess of displacing olefin, it would be anticipated that almost every elimination would result in a displacement. Fortunately, the discrepancy is not real. The rates of isomerization discussed in Chapter 9 refer to the boron-hydrogen catalyzed reaction. In the presence of excess olefin, these catalytic species cannot exist. Consequently, the displacement reaction must measure the approximate rate of the uncatalyzed elimination stage.

This is a fortunate circumstance. By utilizing a slight excess of diborane it is possible rapidly to isomerize the initial organoborane to place the boron atom at the end of a chain or exocyclic to a ring. The addition of the displacing olefin effectively destroys the isomerization catalyst, so that back-isomerization is not important. Consequently, it becomes possible to realize the olefin from the isomerized organoborane.

Contrathermodynamic Isomerization of Olefins

Olefins containing internal double bonds are thermodynamically more stable than those with terminal bonds.

On the other hand, organoboranes with the boron atom in an internal position appear to be less stable than those with the boron atom at the terminal position.

It follows that hydroboration, isomerization of the organoborane, followed by displacement, provides a convenient synthetic route for the contra-thermodynamic transformation of the olefin structure.

For example, the procedure readily permits the conversion of 2-pentene into 1-pentene.

Tertiary olefins, such as 3-ethyl-2-pentene, are highly stabilized relative to primary olefins. Thus treatment of such olefins with acid produces no significant isomerization of the double bond to the terminal position. However, application of the isomerization process provides an 82 per cent yield of product, analyzing for 98 per cent 3-ethyl-1-pentene and only 2 per cent 3-ethyl-2-pentene.

Finally, it has proved possible to achieve the conversion of α-pinene to β-pinene.

It is evident that this contrathermodynamic isomerization of olefins should provide a welcome addition to the repertoire of the organic chemist.

11 | Hydroboration of Hindered Olefins

The initial exploration of the scope of the hydroboration reaction[1] (Chapter 6) revealed that the great majority of olefins react readily under standard conditions, 1 hour at 25°, to yield the corresponding trialkylboranes.

$$3RCH{=}CH_2 + BH_3 \xrightarrow[\text{25°, 1 hr}]{\text{DG}} (RCH_2CH_2)_3B$$

However, in the case of certain highly substituted olefins, such as 2-methyl-2-butene and 2,3-dimethyl-2-butene, the reaction appeared to stop short of this stage, yielding the corresponding dialkylborane and monoalkylborane, respectively.

$$2 \begin{array}{c} H_3C \quad CH_3 \\ | \qquad | \\ C{=}C \\ | \qquad | \\ H_3C \quad H \end{array} + BH_3 \xrightarrow[\text{25°, 1 hr}]{\text{DG}} \begin{array}{c} H_3C \quad CH_3 \\ | \qquad | \\ H{-}C{-}C{-})_2BH \\ | \qquad | \\ H_3C \quad H \end{array}$$

$$\begin{array}{c} H_3C \quad CH_3 \\ | \qquad | \\ C{=}C \\ | \qquad | \\ H_3C \quad CH_3 \end{array} + BH_3 \longrightarrow \begin{array}{c} H_3C \quad CH_3 \\ | \qquad | \\ H{-}C{-}C{-}BH_2 \\ | \qquad | \\ H_3C \quad CH_3 \end{array}$$

[1] H. C. Brown and B. C. Subba Rao, *J. Am. Chem. Soc.*, **81**, 6428 (1959).

Since that time the product from 2-methyl-2-butene, bis-(3-methyl-2-butyl)-borane or disiamylborane has proved of major value as a selective hydroborating[2-4] (Chapter 13) and reducing agent[5] (Chapter 18). Similarly, the trimethylamine addition compound of *t*-butylborane has been utilized as a hydroborating agent.[6]   These demonstrations of the utility of these mono- and dialkylboranes made it desirable to explore in more detail the hydroboration of a number of representative olefins.[7]   It was hoped that information gained would permit definition of conditions for convenient synthesis of a number of mono- and dialkylboranes.[8]

[It is convenient to discuss these compounds as derivatives of borane, BH_3 even though both borane and these alkyl derivatives normally exist as dimers (Chapter 12).   Similarly, it is convenient to discuss the reactions described in this paper in terms of the molecules of olefin reacting under the indicated conditions with each borane (BH_3) or borane equivalent ($\frac{3}{4}NaBH_4 + BF_3$).]

The procedure consisted of mixing the olefin in diglyme (usually 1.5 *M*) with the calculated quantity of sodium borohydride (generally three olefins per borane equivalent).   Hydroboration was achieved by adding boron trifluoride in diglyme to the solution, maintaining the temperature constant at either 0 or 25°.   At appropriate intervals of time, samples were removed and analyzed for residual olefin by gas chromatography.

In a number of cases an alternative procedure was utilized.   The olefin was added to a standard solution of diborane in tetrahydrofuran at 0 or 25°, and the decrease of the olefin with time followed similarly.

Cyclic Olefins

The hydroboration of cyclopentene in diglyme at 0° proceeds rapidly to the formation of the tricyclopentylborane.

[2] H. C. Brown and G. Zweifel, *J. Am. Chem. Soc.*, **83**, 1241 (1961).
[3] G. Zweifel, K. Nagase, and H. C. Brown, *J. Am. Chem. Soc.*, **84**, 190 (1962).
[4] H. C. Brown and A. Moerikofer, *J. Am. Chem. Soc.*, **83**, 3417 (1961).
[5] H. C. Brown and D. B. Bigley, *J. Am. Chem. Soc.*, **83**, 486 (1961).
[6] M. F. Hawthorne, *J. Am. Chem. Soc.*, **83**, 2541 (1961).
[7] H. C. Brown and A. W. Moerikofer, *J. Am. Chem. Soc.*, in press.
[8] H. C. Brown and G. J. Klender, *J. Inorg. Chem.*, in press.

Within a few minutes, the observed ratio of cyclopentene utilized per borane equivalent is essentially quantitative and does not change significantly with time (Figure 11-1).

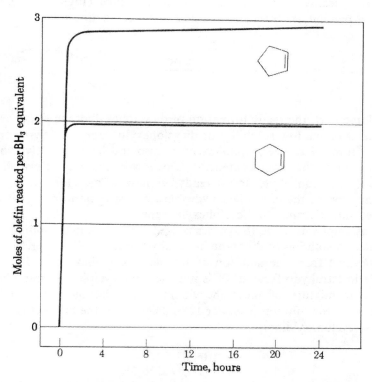

Figure 11-1   The hydroboration of cyclopentene and cyclohexene in diglyme at 0° (three olefins per borane equivalent).

The results indicate that it is not possible by these methods to achieve a simple synthesis of dicyclopentylborane.   An alternative procedure, in which cyclopentene is added to an excess of diborane in tetrahydrofuran (one C_5H_{10} per BH_3), provides a 64 per cent yield of dicyclopentylborane.[9] This procedure is satisfactory for applications in which the concurrent presence of excess diborane and tricyclopentylborane offer no difficulties, such as in the preparation of dicyclopentylborinic acid.[9]   However, the presence of diborane is a major disadvantage in cases in which the product is to be utilized for selective hydroborations[2] or reductions.[5]

[9] H. C. Brown, A. Tsukamoto, and D. B. Bigley, *J. Am. Chem. Soc.*, **82**, 4703 (1960).

In the case of cyclohexene, the reaction at 25° takes place rapidly to the dicyclohexylborane stage, with further reaction to the tricyclohexylborane stage proceeding, but at a much slower rate. Indeed, at 0° the reaction proceeds cleanly to the dicyclohexylborane stage (Figure 11-1). The product precipitates from solution as a white crystalline solid.

In large part, the ready formation of dicyclohexylborane at 0° must be the result of the low solubility of dicyclohexylborane in diglyme at that temperature. However, cyclohexene is also much less reactive toward hydroboration than cyclopentene.[2,4] This lower reactivity of cyclohexene probably also contributes to the ready synthesis of dicyclohexylborane.

In any event, the controlled hydroboration of cyclohexene provides a simple synthetic route to dicyclohexylborane.

An attempt was made to synthesize cyclohexylborane by adding cyclohexene to a solution of diborane in tetrahydrofuran. However, the reaction product from the addition of 1 mole of cyclohexene to 1 mole of borane in tetrahydrofuran at 0° is not the monocyclohexylborane but an equimolar mixture of dicyclohexylborane and borane. The infrared spectrum of the solution (Chapter 12) indicates that the product is present as 1,1-dicyclohexyldiborane.

Acyclic Olefins

At 25° in diglyme, 2-methyl-2-butene (1.5 M) reacts rapidly with sodium borohydride–boron trifluoride (0.5 M in borane equivalent) to form disiamylborane, and the reaction then proceeds at a modest rate toward the formation of trisiamylborane. At 0°, the initial product, disiamylborane, precipitates from solution as a white solid. Reaction proceeds beyond this stage, but at a relatively slow pace — even after 24 hours, the formation of trisiamylborane is only 40 per cent complete (Figure 11-2).

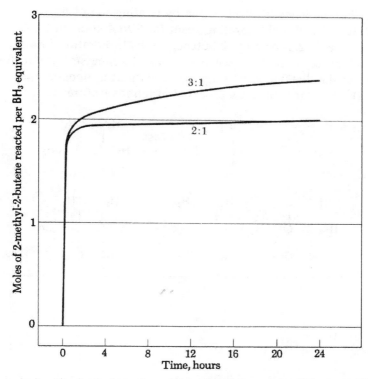

Figure 11-2 The hydroboration of 2-methyl-2-butene in diglyme at 0° (three olefins per borane equivalent and two olefins per borane equivalent).

In the same way, 2-methyl-2-butene in tetrahydrofuran reacts rapidly at 0° to the disiamylborane stage, with subsequent reaction to the trisiamylborane stage proceeding at a much slower pace. In this solvent no precipitation occurs for a solution 0.5 M in borane.

Disiamylborane is readily synthesized by treating the olefin with the calculated quantity of hydroborating agent for 2 to 4 hours at 0°.[2,4]

In the case of 2,3-dimethyl-2-butene, the hydroboration in diglyme at 25 and 0° proceeds rapidly to the monoalkylborane stage, with further reaction to the dialkylborane stage proceeding at a moderate pace (Figure 11-3). Similar results are observed in tetrahydrofuran.

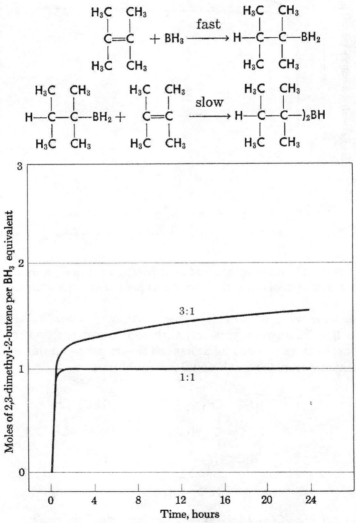

Figure 11-3   The hydroboration of 2,3-dimethyl-2-butene in diglyme solution at 0° (three olefins per borane equivalent and one olefin per borane equivalent).

In these experiments there is present a large excess of olefin. Use of the theoretical quantity of olefin and hydroborating agent results in the direct formation of 2,3-dimethyl-2-butylborane. The latter is proving to be a useful reagent, and the term thexylborane (contracted from *t*-hexylborane) has been proposed as a convenient common name.

2,4,4-Trimethyl-2-pentene is a trisubstituted olefin related to 2-methyl-2-butene, but with larger steric requirements.

The hydroboration of 2,4,4-trimethyl-2-pentene in diglyme at 25° proceeds rapidly to the monoalkylborane stage and quite slowly beyond, approaching the value two olefins per BH_3, only after 2 to 4 days. Treatment of the olefin at 0° with the calculated quantity of hydroborating agent permits the clean synthesis of the monoalkylborane, 2,4,4-trimethyl-3-pentylborane.

Thus, the controlled hydroboration of both 2,3-dimethyl-2-butene and 2,4,4-trimethyl-3-pentene provides quantitative syntheses of the corresponding monoalkylboranes.

Substituted Cyclic Olefins

The hydroboration of 1-methylcyclopentene is rapid to the dialkylborane stage, but proceeds beyond that stage at a slow but measurable rate. The reaction characteristics are very similar to that of 2-methyl-2-butene, so that there should be no difficulty in controlling the hydroboration to produce the dialkyborane, bis-(*trans*-2-methylcyclopentyl)-borane.

The lower reactivity of the cyclohexene structure reveals itself in the behavior of 1-methylcyclohexene; the reaction proceeds rapidly and cleanly to the dialkylborane stage, bis-(*trans*-2-methylcyclohexyl)-borane.

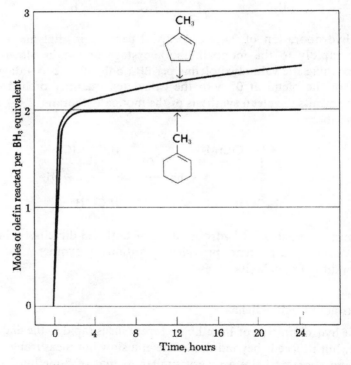

The latter precipitates as a white solid and exhibits no tendency to react further, even in the presence of a large excess of olefin (Figure 11-4).

Figure 11-4 The hydroboration of 1-methylcyclopentene and 1-methyl-cyclohexene in diglyme at 0° (three olefins per borane equivalent).

Bicyclic Olefins

Norbornene reacts rapidly and completely to the trialkylborane stage. The behavior is similar to that of cyclopentene (Figure 11-1).

α-Pinene can be considered a derivative of 1-methylcyclohexene. In diglyme at 25° it is readily converted into diisopinocampheylborane, which precipitates from solution.

The product is much more soluble in tetrahydrofuran. In this solvent the reaction does not appear to proceed to completion unless an excess of α-pinene is present. Removal of the excess results in the slow appearance of small quantities of α-pinene. Consequently, it appears that the reaction involves an equilibrium which is forced to completion in diglyme by the insolubility of the product but which in tetrahydrofuran requires an excess of α-pinene to achieve complete conversion to the dialkylborane.

α-Pinene is available in optically active form. The hydroboration of optically active α-pinene makes available an optically active dialkylborane. This material accomplishes asymmetric hydroborations, leading to optically active boranes, and through oxidation to alcohols with optical purities in the range 80 to 90 per cent[10] (Chapter 14).

The careful addition of α-pinene to diborane in tetrahydrofuran at 0°, in the ratio one α-pinene per BH_3, does not lead to a monoalkylborane. Just as in the case of cyclohexene, described earlier, the reaction appears to proceed predominantly to the 1,1-diisopinocampheyldiborane, $R_2BH_2BH_2$.

---

[10] H. C. Brown and G. Zweifel, *J. Am. Chem. Soc.*, **83**, 486 (1961).

Disubstituted Terminal Olefins

The hydroboration of 2-methylpropene (isobutylene), 2-methyl-1-butene, 2,4,4-trimethyl-1-pentene, and β-pinene was examined to ascertain the possibility of achieving a convenient synthesis of one or more dialkylboranes with olefins of this structure.   Treatment of 2,4,4-trimethyl-1-pentene or β-pinene with the hydroborating agent leads to the immediate uptake of three olefins per BH_3, as observed for cyclopentene (Fig. 11-1).

Addition of the olefin to the calculated quantity of borane in tetrahydrofuran leads to the formation of the dialkylborane in yields of 60 to 70 per cent.   No significant quantities of the monoalkylborane are formed (Table 11-1).

Table 11-1

$$2\;\underset{R}{\overset{R}{>}}C{=}CH_2 + BH_3 \xrightarrow{\;THF,\;0°\;} BH_3 + R_2BH + R_3B$$

2-Methylpropene	12%	64%	24%
2-Methyl-1-butene	11	53	24
2,4,4-Trimethyl-1-pentene	10	70	20
β-Pinene	13	61	26

It is evident that this procedure provides reasonable yields of the dialkylborane, provided the byproducts (residual borane and trialkylborane) can either be tolerated or separated.

Hydroboration Mechanism

In tetrahydrofuran solution, diborane exists as the monomeric species, tetrahydrofuran-borane.[11,12]   On the other hand, both the monoalkylboranes and the dialkylboranes are dimeric in tetrahydrofuran[8] (Chapter 12).

In this solvent it is presumed that the first step in the hydroboration reaction must involve the addition of the borane particle to the olefin to

[11] J. R. Elliott, W. L. Roth, G. F. Roedel, and E. M. Boldebuck, *J. Am. Chem. Soc.*, **74**, 5211 (1952).

[12] B. Rice, J. A. Livasy, and G. W. Schaeffer, *J. Am. Chem. Soc.*, **77**, 2750 (1955).

give the corresponding monoalkylborane. This addition presumably proceeds through the four-center transition state discussed earlier (Chapter 8).

$$\underset{|}{\overset{|}{C}}=\underset{|}{\overset{|}{C}} + H_3B : THF \longrightarrow H-\underset{|}{\overset{|}{C}}-\underset{|}{\overset{|}{C}}-BH_2 + THF$$

In the case of highly hindered olefins, such as 2,4,4-trimethyl-2-pentene and 2,3-dimethyl-2-butene, the reaction to place an additional alkyl group upon the same boron atom becomes relatively slow, and the monoalkylborane particles dimerize to form the *sym*-dialkyldiboranes, which are identified in the reaction solution and isolated therefrom (Chapter 12).

On the other hand, it appears that the reaction of a less hindered olefin with the monoalkylborane intermediate is faster than its reaction with tetrahydrofuran-borane. As a result, with these olefins the monoalkylborane intermediates are converted into dialkylborane derivatives, in spite of the presence of considerable quantities of tetrahydrofuran-borane.

$$\underset{|}{\overset{|}{C}}=\underset{|}{\overset{|}{C}} + RBH_2 \longrightarrow R_2BH$$

In order to account for the identification of the 1,1-dialkyldiborane species in the solution, the 1,1-dialkylboranes must be capable of competing successfully with tetrahydrofuran for the borane groups.

$$R_2BH + H_3B : THF \rightleftharpoons \quad + THF$$

The further addition of olefin converts the 1,1-dialkyldiborane derivatives into the corresponding tetraalkyldiborane, and, in the case of less hindered derivatives, into trialkylboranes.

In the case of the disubstituted terminal olefins (2-methylpropene, 2-methyl-1-butene, 2,4,4-trimethyl-2-pentene, and β-pinene), the reaction of olefin with the tetraalkyldiborane appears to proceed at a rate competitive with the reaction of the olefin with tetrahydrofuran-borane. As a result, the reaction contains both the borane and trialkylborane in addition to the 60 to 70 per cent of the desired tetraalkyldiborane.

Conclusion

The observations here summarized on the hydroboration of hindered olefins demonstrate that the reaction in many cases can be controlled to provide an essentially quantitative yield of a particular monoalkyl- or dialkylborane. The reaction has been utilized to synthesize a number of such derivatives, including disiamylborane, thexylborane, 2,4,4-trimethyl-3-pentylborane, dicyclohexylborane, bis-(*trans*-2-methylcyclopentyl)-borane, bis-(*trans*-2-methylcyclohexyl)-borane, and diisopinocampheylborane.

The physical and chemical properties of several of these derivatives are described in Chapter 12. Several of these derivatives exhibit very useful characteristics as selective hydroborating and reducing agents. These characteristics will be summarized in Chapters 13, 14, and 18.

12 | Alkylboranes

In the great majority of olefins hydroboration proceeds quantitatively to the formation of the trialkylborane[1] (Chapter 16).

$$3 \quad \underset{\underset{H}{|}}{\overset{\overset{H_3C}{|}}{C}} = \underset{\underset{H}{|}}{\overset{\overset{CH_3}{|}}{C}} \quad + BH_3 \longrightarrow H - \underset{\underset{H}{|}}{\overset{\overset{H_3C}{|}}{C}} - \underset{\underset{H}{|}}{\overset{\overset{CH_3}{|}}{C}} -)_3B$$

However, in the case of more hindered olefins, such as 2-methyl-2-butene, cyclohexene, 1-methylcyclopentene, 1-methylcyclohexene, and α-pinene, it was observed that the reaction could be directed cleanly to the formation of the dialkylborane[1,2] (Chapter 11).

$$2 \quad \underset{\underset{H_3C}{|}}{\overset{\overset{H_3C}{|}}{C}} = \underset{\underset{H}{|}}{\overset{\overset{CH_3}{|}}{C}} \quad + BH_3 \longrightarrow H - \underset{\underset{H_3C}{|}}{\overset{\overset{H_3C}{|}}{C}} - \underset{\underset{H}{|}}{\overset{\overset{CH_3}{|}}{C}} -)_2BH$$

Finally, with even more hindered olefins, such as 2,3-dimethyl-2-butene and 2,2,4-trimethyl-2-pentene, the reaction can be utilized for the quantitative synthesis of the monoalkylborane[1,2] (Chapter 11).

$$\underset{\underset{H_3C}{|}}{\overset{\overset{H_3C}{|}}{C}} = \underset{\underset{CH_3}{|}}{\overset{\overset{CH_3}{|}}{C}} \quad + BH_3 \longrightarrow H - \underset{\underset{H_3C}{|}}{\overset{\overset{H_3C}{|}}{C}} - \underset{\underset{CH_3}{|}}{\overset{\overset{CH_3}{|}}{C}} - BH_2$$

[1] H. C. Brown and B. C. Subba Rao, *J. Am. Chem. Soc.*, **81**, 6428 (1959).
[2] H. C. Brown and A. W. Moerikofer, *J. Am. Chem. Soc.*, in press.

This synthesis makes a number of mono- and dialkylboranes readily available as pure substances. Some of these derivatives are proving to be of considerable value as selective hydroborating[3,4] (Chapters 13 and 14) and reducing agents[5] (Chapter 18). They are also of interest as representative intermediates in the hydroboration reaction. However, very little has been done with such derivatives and there is very little known about their chemistry. It appeared desirable to undertake the isolation and characterization of a number of such monoalkyl- and dialkylboranes.

This investigation has demonstrated that these substances exist as dimers, even in ether solvents. Thus they are alkyldiboranes. However, it is often convenient to discuss them in terms of the monomeric formulation, and to consider them as diborane derivatives only in cases where the diborane structure is of significance in the chemistry and properties of the product.

Alternative Syntheses

In the past the alkylboranes were available primarily through the redistribution reaction between diborane and trialkylboranes, discovered and applied by Schlesinger and his co-workers. For example, diborane and trimethylborane react, slowly at room temperature and rapidly at 70 to 80°, to give a mixture of methyldiboranes.[6] The products isolated by fractionation and identified were monomethyldiborane, 1,1-dimethyldiborane, trimethyldiborane, and sym-tetramethyldiborane.

$$B_2H_6 + B(CH_3)_3$$

[3] H. C. Brown and G. Zweifel, *J. Am. Chem. Soc.*, **83**, 1241 (1961).

[4] H. C. Brown and G. Zweifel, *J. Am. Chem. Soc.*, **83**, 486 (1961).

[5] H. C. Brown and D. B. Bigley, *J. Am. Chem. Soc.*, **83**, 486 (1961).

[6] H. I. Schlesinger and A. O. Walker, *J. Am. Chem. Soc.*, **57**, 621 (1935).

Similar products were realized from the reaction of diborane and both triethylborane and tri-*n*-propylborane.[7]

Later it proved possible to synthesize *sym*-dimethyldiborane by treating monomethyldiborane with a weak base capable of reacting with borane but not with methylborane.[8] The postulated reaction course is as indicated.

In theory the symmetrical dimethyl derivative should be capable of existing as a pair of *cis-trans* isomers, but such isomers have not yet been characterized.

It was reported that the individual compounds are stable at low temperatures ($-80°$) but have only limited stability at room temperature, quickly undergoing redistribution to an equilibrium distribution. The products react with trimethylamine to form addition compounds which are much more stable.[8]

$$(CH_3BH_2)_2 + 2(CH_3)_3N \longrightarrow 2(CH_3)_3N : BH_2CH_3$$

$$[(CH_3)_2BH]_2 + 2(CH_3)_3N \longrightarrow 2(CH_3)_3N : BH(CH_3)_2$$

They also undergo hydrolysis and methanolysis, reactions that are highly useful in the identification of the structures.[6]

$$(CH_3)_2BH_2BH_2 + 4CH_3OH \longrightarrow (CH_3)_2BOCH_3 + (CH_3O)_3B + 4H_2$$

$$CH_3HBH_2BHCH_3 + 4CH_3OH \longrightarrow 2CH_3B(OCH_3)_2 + 4H_2$$

[7] H. I. Schlesinger, L. Horvitz, and A. B. Burg, *J. Am. Chem. Soc.*, **58**, 407 (1936).

[8] H. I. Schlesinger, H. W. Flodin, and A. B. Burg, *J. Am. Chem. Soc.*, **61**, 1078 (1939).

In the absence of any residual boron-hydrogen bonds, the esters are quite stable to disproportionation[9] and can be isolated and identified.

Recently, it was observed that the synthesis of *sym*-diethyldiborane from the monoethyl derivative proceeds spontaneously at room temperature.[10] Apparently no base is required, as indicated in the related synthesis of the methyl derivative.[8] The reaction appears to involve a simple exchange of monoethylborane and borane groups. By removing the diborane the equilibrium can be shifted quantitatively to the formation of *sym*-diethyldiborane, and the latter exhibits considerable stability at room temperature (no observed change in 24 hours).

$$2C_2H_5HBH_2BH_2 \rightleftharpoons (C_2H_5BH_2)_2 + (BH_3)_2$$

The reaction of diborane with triphenylborane at 80° and 2.2 atm has recently been utilized also to achieve the synthesis of *sym*-diphenyldiborane, $C_6H_5HBH_2BHC_6H_5$, m.p. 85°.[11]

The equilibration of tri-*n*-pentylborane and tricyclopentylborane (from the hydroboration of the corresponding olefins) with diborane in tetrahydrofuran solution proceeds at a reasonable pace at 25°, more rapidly at 50 to 55° (Figures 12-1 and 12-2). Evidently this equilibration can be controlled to provide a convenient route to the monoalkyl- or dialkylboranes.[12] However, this reaction involves excess diborane, some of which remains. Consequently, this procedure is satisfactory only in cases in which the presence of the excess diborane is not a problem. Such is the case where the reaction product is methanolized to produce methyl borate and the esters of borinic and boronic acids, readily separable by distillation.[12]

The reduction of the borinic or boronic esters and related derivatives with lithium aluminum hydride provides an alternative synthetic route to these derivatives. For example, Wiberg and his co-workers also obtained *sym*-diphenyldiborane by the reduction of phenylboron dichloride.[11] Similarly, Hawthorne achieved the synthesis of trimethylamine-*t*-butylborane by reducing *t*-butylboroxine (from *t*-butylmagnesium chloride and

[9] R. Köster, *Angew. Chem.*, **73**, 66 (1961).

[10] I. J. Solomon, M. J. Klein, and K. Hattori, *J. Am. Chem. Soc.*, **80**, 4520 (1958).

[11] E. Wiberg, J. E. F. Evans, and H. Nöth, *Z. Naturforsch.*, **13b**, 263 (1958).

[12] H. C. Brown, A. Tsukamoto, and D. B. Bigley, *J. Am. Chem. Soc.*, **82**, 4703 (1960).

methylborate at low temperatures) with lithium aluminum hydride in the presence of trimethylamine.[13]

$$(CH_3)_3CMgCl \xrightarrow{(CH_3O)_3B} (CH_3)_3CB(OCH_3)_2 \xrightarrow[\Delta]{H_2O}$$

$$[(CH_3)_3CBO]_3 \xrightarrow[(CH_3)_3N]{LiAlH_4} (CH_3)_3CBH_2 : N(CH_3)_3$$

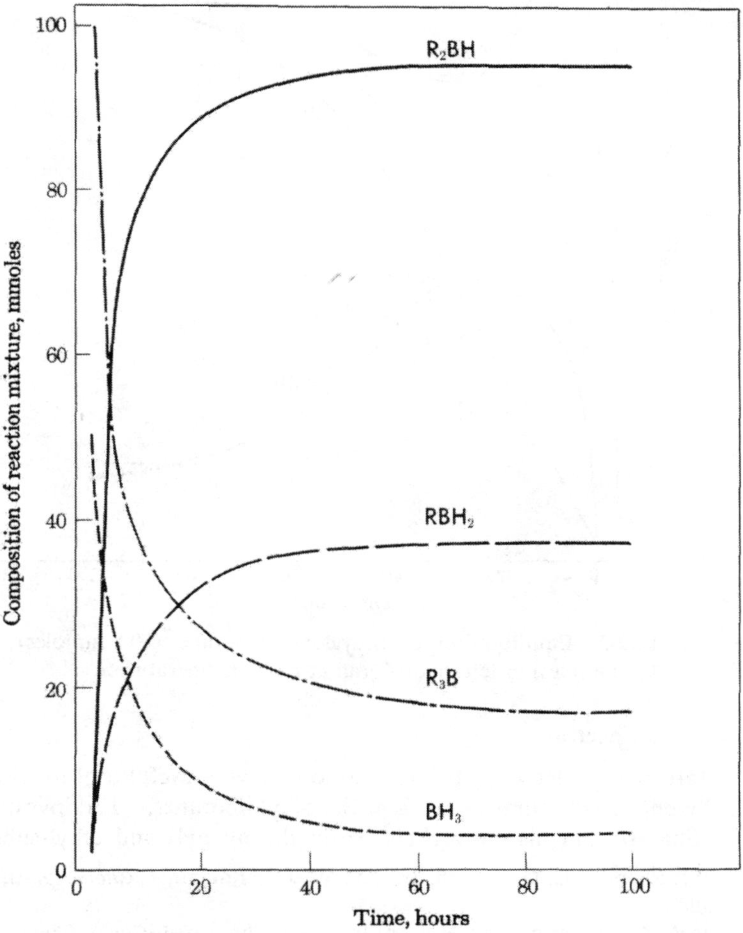

Figure 12-1 Equilibration of tri-n-pentylborane (100 mmoles) and diborane (25 mmoles) in tetrahydrofuran at room temperature.

[13] M. F. Hawthorne, *J. Am. Chem. Soc.*, **81**, 5836 (1959).

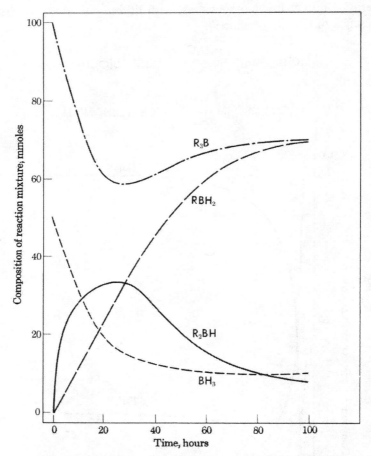

Figure 12-2 Equilibration of tricyclopentylborane (100 mmoles) and diborane (25 mmoles) in tetrahydrofuran at room temperature.

Infrared Spectra

Infrared spectroscopy has proved to be a very useful tool for the establishment of structural features in the alkyldiboranes. The investigations of Shapiro and his co-workers[14,15] on the methyl- and ethyl-substituted

[14] I. Shapiro, C. O. Wilson, Jr., and W. J. Lehmann, *J. Chem. Phys.*, **29,** 237 (1958).

[15] W. J. Lehmann, C. O. Wilson, Jr., and I. Shapiro, *J. Chem. Phys.*, **32,** 1088, 1786 (1960); **33,** 590 (1960).

diboranes in the gas phase have established that the presence of alkyl substituents greatly simplifies the boron-hydrogen stretching spectrum over that observed in diborane itself. They also point out that certain bands appear to be characteristic for different alkyldiborane structures.

For example, of the four fundamental stretching modes possible for the diborane hydrogen bridge, only one appears in the alkyldiboranes, a strong band located in the 1600 to 1500 cm^{-1} (6.2 to 6.7 μ) region. Consequently, the presence of a strong band in this region indicates that a given alkylborane exists as the dimer.

The terminal boron-hydrogen stretching modes are also characteristic, appearing in the 2600 to 2500 cm^{-1} (3.8 to 4.0 μ) region. Moreover, differences in the bands in this region are characteristic, and can be utilized to distinguish between different alkyldiboranes. The tetraalkyldiboranes, $R_2BH_2BR_2$, have no terminal hydrogens and are characterized by the absence of absorption in this region. The *sym*-dialkyldiboranes, $RHBH_2BHR$, and the trialkyldiboranes, R_2BH_2BHR, exhibit a single, strong stretching frequency in this region, usually at 2500 cm^{-1} (4.0 μ), characteristic of the structure $\rangle$BHR. The *unsym*-dialkyldiborane, $R_2BH_2BH_2$, exhibits a double absorption in the 2600 to 2500 cm^{-1} (3.8 to 4.0 μ) region, arising from the asymmetric and symmetric stretching vibrations of the terminal $\rangle$BH$_2$ group. The monoalkyldiborane, $RHBH_2BH_2$, exhibits both the band at 2500 cm^{-1} (4.0 μ) due to the $\rangle$BHR structure and the double absorption at 2600 to 2500 cm^{-1} (3.8 to 4.0 μ) due to the terminal $\rangle$BH$_2$. Consequently, in the case of this compound, the two bands in the 2600 to 2500 cm^{-1} (3.8 to 4.0 μ) region are characterized by considerably different intensities, that at $\sim$ 3.8 μ being considerably less than the symmetric band at 4.0 μ, whereas the double absorption bands of the *unsym*-dialkyldiboranes are of equal intensity.

The unsymmetrical dialkyldiborane, $R_2BH_2BH_2$, also exhibits a unique change in the hydrogen bridge band. It shifts from the usual value at 1600 cm^{-1} (6.2 μ) to a lower frequency, 1550 to 1540 cm^{-1} (6.6 μ), and another of the bridge modes becomes prominent around 2100 cm^{-1} (4.8 μ). The correlations for methyl and ethyldiboranes are summarized in Table 12-1.

In the conversion of diborane and its derivatives to a simple addition compound, such as trimethylamine-borane, major changes occur in the spectrum.[16] Not only do the bridge bands disappear in this compound, but the terminal asymmetric and symmetric absorptions are shifted to lower frequencies, 2400 to 2200 cm^{-1} (4.2 to 4.5 μ). Furthermore, a

---

[16] B. Rice, R. J. Galiano, and W. J. Lehmann, *J. Phys. Chem.*, **61**, 1222 (1957).

Table 12-1

Correlation of Characteristic Boron-Hydrogen Stretching Vibrations with Structure of the Methyl- and Ethyl-Substituted Diboranes

Compound	Position,[a] cm^{-1}	Type of band	Mode	Characteristic unit
$MeHBH_2BH_2$ $Me_2BH_2BH_2$	2571 (2597)[b] 2513 (1419)[b]	Strong, sharp doublet	B—H stretching (asymmetric, symmetric)	
$MeHBH_2BHMe$ Me_2BH_2BHMe	2519 (2506)[c]	Strong, sharp singlet	B—H stretching	
$MeHBH_2BH_2$	2100 (2080)	Medium, broad band	B—H bridge, symmetric in phase	
MeB_2H_5 to $Me_2BH_2BMe_2$	1592 (1593)[b]	Broad, strong band	B—H bridge, asymmetric in phase	
$Me_2BH_2BH_2$	1540 (1550)	Very strong, sharp band	B—H bridge, and terminal	

[a] Values in parenthesis for ethyl derivatives.
[b] Values reported for MeB_2H_5 and EtB_2H_5 only.
[c] Values reported for symmetrical dialkyldiboranes only.

180

strong BH_3 asymmetric deformation appears at 1170 cm^{-1} (8.6 μ), and, as a consequence, an overtone band appears between the asymmetric and symmetric stretching modes. Therefore, the resultant spectrum in the boron-hydrogen stretching region appears as a strong band at 2400 cm^{-1} (4.2 μ) with two strong shoulders about 0.1 μ apart, representing the BH_3 overtone and the symmetric B—H stretching vibration, respectively.

sym-Tetrasiamyldiborane

Disiamylborane was synthesized in ethyl ether by the quantitative reaction of 4 moles of 2-methyl-2-butene with 1 mole of diborane. The product, a white solid, exhibits a molecular weight (from vapor-pressure lowering) of 315 in ethyl ether and 297 in tetrahydrofuran. Therefore, the molecule exists as the dimer in both of these solvents (calculated molecular weight of the dimer, 308).

The infrared spectrum of tetrasiamyldiborane in cyclohexane solution exhibits a strong absorption at 1565 cm^{-1} (6.39 μ), characteristic of the bridge hydrogen absorption. The spectrum in tetrahydrofuran solution is similar, with the boron-hydrogen bridge absorption undergoing a slight shift to 1551 cm^{-1} (Figure 12-3).

Freshly synthesized samples melted at approximately 40 to 44°. However, a sample once melted, frequently remained all or partially liquid. The observations are indicative of some kind of irreversible change, associated with the unusual structure of the dimer. The hydroboration introduces an asymmetric center in each siamyl group, so that the product presumably is an isomeric mixture of five different diastereoisomers, 3 *meso* and 2 *dl*.

Presumably, these diastereoisomers interconvert relatively easily, by exchange of borane groupings, and cause the observed melting point to be both indefinite and dependent on the history of the sample.

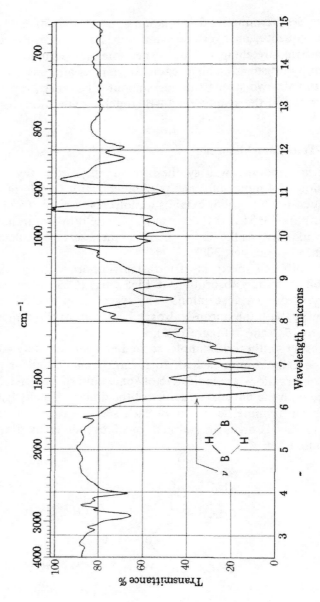

Figure 12-3 Infrared spectrum of *sym*-tetrasiamyldiborane.

182

A solution of tetrasiamyldiborane in diglyme rapidly absorbs an equivalent quantity of diborane at 0°, the reaction proceeding 83 per cent toward completion.

$$(\text{Sia}_2\text{BH})_2 + (\text{BH}_3)_2 \xrightarrow[0°]{\text{DG}} 2\text{Sia}_2\text{BH}_2\text{BH}_2$$

The formation of the unsymmetrical disiamyldiborane indicated above is supported by the infrared spectrum (Figure 12-4). This reveals a strong doublet at 2571 cm^{-1} (3.89 μ) and 2500 cm^{-1} (4.0 μ) of equal intensity, characteristic of 1,1-dialkyldiboranes. Moreover, there is present a very strong absorption, in the bridge region, at 1517 cm^{-1} (6.39 μ).

Evidently the equilibrium observed by Solomon et al.[10] in the *sym*-ethyldiborane system is much more favorable in the present case. Possibly the large steric strain in tetrasiamyldiborane is relieved in forming the *unsym*-disiamyldiborane and provides a powerful driving force favoring the transformation.

Trimethylamine reacts with tetrasiamyldiborane to form an unstable 1:1 addition compound.

$$(\text{Sia}_2\text{BH})_2 + 2(\text{CH}_3)_3\text{N} \overset{0°}{\rightleftharpoons} 2(\text{CH}_3)_3\text{N}:\text{BHSia}_2$$

The product is a liquid at $-78.5°$ (m.p. -86 to $-82°$), with a dissociation pressure of amine above the addition compound of 35.2 mm at 0°. In cyclohexane solution the product exhibits a very broad, rather weak absorption at 2366 cm^{-1} (4.3 μ) and a bridge absorption of reduced intensity at 1546 cm^{-1} (6.47 μ), indicating considerable dissociation in solution at 25°. Trimethylamine-dimethylborane is quite stable at room temperature. Apparently, the larger steric requirements of the siamyl groups must be responsible for the enhanced tendency toward dissociation.

Tetrasiamylborane reacts rapidly with methanol at 0° to form the ester. The latter reacts with dry air to undergo a slow oxidation.

$$(\text{Sia}_2\text{BH})_2 + 2\text{CH}_3\text{OH} \longrightarrow 2\text{Sia}_2\text{BOCH}_3 + 2\text{H}_2$$

sym-Dithexyldiborane

2,3-Dimethyl-2-butene is converted quantitatively into the corresponding borane by treatment with the theoretical quantity of diborane in

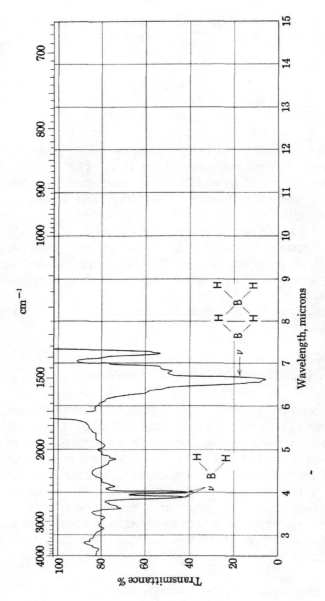

Figure 12-4   Infrared spectrum of *unsym*-disiamyldiborane, produced by the reaction of diborane with *sym*-tetrasiamyldiborane at 0°.

184

diethyl ether at 0°. The product is a water-clear liquid, which exhibits a molecular weight (vapor-pressure lowering) of 196 in ethyl ether and 193 in tetrahydrofuran. The value calculated for the dimer is 197.

The infrared spectra in cyclohexane and in tetrahydrofuran were identical. The bridge-hydrogen band was observed at 1565 cm^{-1} (6.39 μ), together with a single strong band at 2540 cm^{-1} (3.98 μ) characteristic of the terminal $>$BHR structure (Figure 12-5).

The data are therefore consistent with the formulation of thexylborane as the simple dimer, *sym*-dithexyldiborane. The product melts relatively sharply and quite reproducibly at −34.7 to −32.5°. The difference in behavior between the siamyl and thexyl derivatives presumably arises from the absence of an asymmetric center in *sym*-dithexyldiborane.

Diborane reacts reversibly at 0° with *sym*-dithexyldiborane to form monothexyldiborane. However, the reaction is less favorable than in the case of *sym*-tetrasiamyldiborane. Under identical conditions, where the reaction with the siamyl derivative proceeds 83 per cent to completion, the corresponding reaction with *sym*-dithexyldiborane proceeds only 43 per cent to completion. At low temperatures it is possible to realize essentially complete conversion, but diborane is evolved at higher temperatures.

$$(t\text{-HexBH}_2)_2 + (BH_3)_2 \underset{0°}{\overset{DG}{\rightleftharpoons}} 2t\text{-HexHBH}_2BH_2$$

sym-Dithexylborane reacts with trimethylamine to form a stable 1:1 addition compound, related to the product synthesized by Hawthorne, trimethylamine-*t*-butylborane.[13]

$$(t\text{-HexBH}_2)_2 + 2(CH_3)_3N \longrightarrow 2(CH_3)_3N : BH_2t\text{-Hex}$$

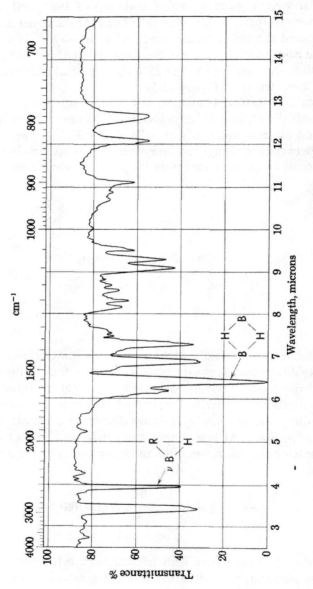

Figure 12-5 Infrared spectrum of *sym*-dithexyldiborane.

186

The product is a white solid at low temperatures, melting to a clear liquid at −3 to −2° at room temperatures. It can be volatilized under vacuum without loss of the amine.

The infrared spectrum shows the characteristic absorption band at 2381 cm^{-1} (4.2 μ) and 2326 cm^{-1} (4.3 μ) for the symmetrical and asymmetrical boron-hydrogen stretching modes of a borane-addition compound.   Moreover, the overtone band observed in this region in trimethylamine-borane, ascribed to a BH$_3$ deformation,[16] is absent.

Methanol reacts rapidly at 0° with *sym*-dithexyldiborane in tetrahydrofuran to form the corresponding boronic ester, stable to dry air at room temperature.

$$(t\text{-HexBH}_2)_2 + 4\text{CH}_3\text{OH} \longrightarrow 2t\text{-HexB(OCH}_3)_2 + 4\text{H}_2$$

sym-Bis-(2,4,4-trimethyl-3-pentyl)-diborane

The hydroboration of 2,4,4-trimethyl-2-pentene proceeds quantitatively to the corresponding borane. The infrared spectrum (Figure 12-6) and molecular weight values confirm that this compound also is dimeric in ether and tetrahydrofuran. The product is a white solid which exhibits indefinite melting characteristics reminiscent of the behavior of tetrasiamyldiborane. As in the case of the latter product, this is attributed to the formation of easily interconvertible diastereoisomers.

sym-Tetracyclohexyldiborane

This product is a white solid, m.p. 103 to 105°, of very low solubility. Its structure was established by observations of the infrared spectra of solutions prepared at 40 to 50°. The presence of a strong bridge band at 1550 cm^{-1} (6.46 μ) with no absorption in the terminal boron-hydrogen region establishes the fact that the product exists as a dimer, both in cyclohexane and tetrahydrofuran solution.

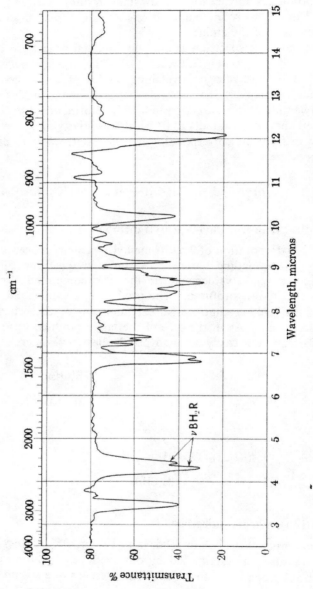

Figure 12-6 Infrared spectrum of trimethylamine-thexylborane.

188

sym-Tetrakis-(*trans*-2-methylcyclohexyl)-diborane

In this case the product is also a white solid of low solubility. Its melting point is indefinite and irreversible, probably due to the presence of diastereoisomers, as discussed earlier. However, the high m.p. range, 116 to 119°, may also result in partial isomerization by migration of the boron. Infrared examination of a suspension reveals that the product exists as the dimer.

sym-Tetraisopinocampheyldiborane

In diglyme α-pinene reacts to form *sym*-tetraisopinocampheyldiborane, the product precipitating as a white solid from solution. However, in tetrahydrofuran, a solvent in which the product is much more soluble, the reaction fails to go to completion. Infrared examination reveals the usual bridge band, but also a distinct sharp singlet at 4.0 μ, characteristic of the terminal BHR structure. This band disappears if excess α-pinene is added and reappears if the excess is removed. Evidently, in this case there is a reversible equilibrium.

The product is an optically active derivative. It was of interest to examine its optical rotation and that of some of its derivatives. Accordingly, d-α-pinene ($[\alpha]_D^{20} +47.5°$) was converted to tetraisopinocampheyldiborane in presence of excess olefin. Corrected for the excess, the rotation of tetraisopinocampheyldiborane is $[\alpha]_D^{20} -37.1°$, $[M]_D^{20} -212.7°$. The methyl ester exhibits the rotation, $[\alpha]_D^{20} -53.6°$, $[M]_D^{20} -169.3°$. Similarly β-pinene ($[\alpha]_D^{20} -21.3°$) yields tri-(*cis*-myrtanyl)-borane, $[\alpha]_D^{20} -27.4°$, $[M]_D^{20} -116°$, and estimated for tetra-(*cis*-myrtanyl)-diborane, $[\alpha]_D^{20} -34.7°$, $[M]_D^{20} -197°$.

The Dimerization Phenomenon

The data reveal that all the alkyldiboranes exist as dimers, even in tetrahydrofuran solution. Even the presence of large bulky alkyl groups, as in *sym*-tetrasiamyldiborane, does not lead to any observable dissociation.

Diborane itself is dimeric, but it dissociates to the monomer in tetra-hydrofuran solution.[17,18] The failure of tetrahydrofuran to dissociate *sym*-dithexyldiborane and related derivatives, as it does diborane itself, is presumably due to the combined polar and steric influences of the alkyl substituent, which reduce the acceptor ability of the boron atom. This causes a decrease in the stability of the alkylborane addition compound relative to that formed by borane itself. On the other hand, the formation of a diborane bridge involves both the acceptor abilities of boron and the donor properties of the hydrogen. Although the alkyl substituent weakens the acceptor properties of the boron atom, it must simultaneously increase the donor properties of the hydrogen atom. The two effects should tend to compensate for each other, so that the bridge structure in the alkyl-substituted diboranes may possess stabilities comparable to that of diborane itself.

These observations on the relative ability of tetrahydrofuran to disso-ciate diborane as compared to its alkyl derivatives are reflected in the data reported for the much stronger base, trimethylamine. This base disso-ciates both diborane and *sym*-dithexyldiborane completely, but the reac-tion with *sym*-tetrasiamyldiborane is an equilibrium, easily reversed.

The fact that the alkylboranes exist as dimers in ether solvents has im-portant consequences in the interpretation of the reactions. It provides an explanation for the effect of boron-hydrogen moieties on the rate of isomerization (Chapter 9), it accounts for the effect of the steric require-ments of the olefin on the products realized in the partial alkylation of diborane (Chapter 11), and it accounts for the observed kinetics of the reaction of tetrasiamyldiborane with olefins (Chapter 13).

[17] J. R. Elliott, W. L. Roth, G. F. Roedel, and E. M. Boldebuck, *J. Am. Chem. Soc.*, **74**, 5211 (1952).

[18] B. Rice, J. A. Livasy, and G. W. Schaeffer, *J. Am. Chem. Soc.*, **77**, 2750 (1955).

13 | Selective Hydroboration with Disiamylborane

It was previously pointed out that the hydroboration of 2-methyl-2-butene proceeds rapidly to the dialkylborane stage, with further reaction to the trialkylborane stage being either slow or negligible under the reaction conditions[1,2] (Chapter 11).

$$2 \quad \underset{H_3C}{\overset{H_3C}{\underset{|}{\overset{|}{C}}}} = \underset{H}{\overset{CH_3}{\underset{|}{\overset{|}{C}}}} + BH_3 \xrightarrow{\text{fast}} H - \underset{H_3C}{\overset{H_3C}{\underset{|}{\overset{|}{C}}}} - \underset{H}{\overset{CH_3}{\underset{|}{\overset{|}{C}}}} -)_2BH$$

$$H - \underset{H_3C}{\overset{H_3C}{\underset{|}{\overset{|}{C}}}} - \underset{H}{\overset{CH_3}{\underset{|}{\overset{|}{C}}}} -)_2BH + \underset{H_3C}{\overset{H_3C}{\underset{|}{\overset{|}{C}}}} = \underset{H}{\overset{CH_3}{\underset{|}{\overset{|}{C}}}} \xrightarrow{\text{very slow}} H - \underset{H_3C}{\overset{H_3C}{\underset{|}{\overset{|}{C}}}} - \underset{H}{\overset{CH_3}{\underset{|}{\overset{|}{C}}}} -)_3B$$

The slowness of the last stage is presumably due to the large steric requirements of both the disiamylborane and the trisubstituted olefin. [The name siamyl has been suggested as a convenient term for the group, $(CH_3)_2CH\dot{C}H(CH_3)$, contracted from "sec-isoamyl."] The large steric requirements of the reagent proved quite useful in achieving the clean monohydroboration of terminal acetylenes[3] (Chapter 16). Accordingly, an exploratory study was undertaken of the utility of this reagent in

[1] H. C. Brown and B. C. Subba Rao, *J. Am. Chem. Soc.*, **81**, 6428 (1959).

[2] H. C. Brown and A. W. Moerikofer, *J. Am. Chem. Soc.*, in press.

[3] H. C. Brown and G. Zweifel, *J. Am. Chem. Soc.*, **81**, 1512 (1959).

achieving some measure of steric control over both the direction and selectivity of hydroboration.[4,5]

Directive Effects

As was discussed earlier (Chapter 7), the hydroboration of terminal olefins, such as 1-hexene, proceeds to place 94 per cent of the boron on the terminal position and 6 per cent on the adjacent secondary.[6] With disiamylborane 99 per cent of the boron adds to the terminal position, only 1 per cent in the secondary position.[5] [The product actually exists as a dimer (Chapter 12). However, it is convenient to discuss it in terms of the monomer, except in cases where the dimeric structure becomes important in the phenomena under consideration.]

$$CH_3CH_2CH_2CH_2CH{=}CH_2 \qquad CH_3CH_2CH_2CH_2CH{=}CH_2$$

$$6\% \ \ 94\% \qquad\qquad 1\% \ \ 99\%$$

$$BH_3 \qquad\qquad\qquad Sia_2BH$$

In the case of styrene, diborane reacts to place 20 per cent of the boron on the secondary position. This drops to 9 per cent with *p*-methoxystyrene. Disiamylborane reduces the addition in the secondary position in these two cases to 2 per cent.

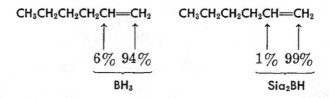

$$20\% \ \ 80\% \qquad\qquad 2\% \ \ 98\%$$

$$BH_3 \qquad\qquad\qquad Sia_2BH$$

[4] H. C. Brown and G. Zweifel, *J. Am. Chem. Soc.*, **82**, 3222 (1960).
[5] H. C. Brown and G. Zweifel, *J. Am. Chem. Soc.*, **83**, 1241 (1961).
[6] H. C. Brown and G. Zweifel, *J. Am. Chem. Soc.*, **82**, 4708 (1960).

The hydroboration of *trans*-4-methyl-2-pentene with diborane reveals very little discrimination between the two positions of the double bond. Disiamylborane exhibits an overwhelming preference for the less-hindered position.

In applying the reagent to several cyclic olefins, such as 3-methylcyclopentene, 3-methylcyclohexene, and 3,3-dimethylcyclohexene, there was observed only a modest control over the direction of hydroboration. For example, the latter olefin provides a 50:50 distribution with diborane, a 40:60 distribution with disiamylborane.[7]

A striking example of the use of the reagent to achieve steric control of the direction of hydroboration is reported by Sondheimer and Nussim.[8] The usual hydroboration of Δ^1-cholestene yields 35 per cent of cholestan-1α-ol and 40 per cent of cholestan-2α-ol. On the other hand, disiamylborane yields 75 per cent of cholestan-2α-ol, the less-hindered isomer, with no detectable amount of cholestan-1α-ol.

Relative Reactivities

Diborane exhibits a mild selectivity between olefins of different structures. For example, treatment of a mixture of 1- and 2-hexene results in

[7] H. C. Brown and G. Zweifel, *J. Am. Chem. Soc.*, **83**, 2544 (1961).
[8] F. Sondheimer and M. Nussim, *J. Org. Chem.*, **26**, 630 (1961).

the preferential reaction of the terminal olefin.[9] However, a multifunctional reagent, such as diborane, possesses an inherent disadvantage for selective reactions. In the case of such a molecule, each successive alkylation stage produces a new reagent with different characteristics. Disiamylborane, with only one active center on the boron atom, appeared more suitable for achieving highly selective reactions.

It was pointed out that the final stage of the reaction of 3-methyl-2-butene with disiamylborane is very slow, so slow that it is possible to follow the uptake of olefin by the reagent. It appeared that this procedure might allow a direct measurement of the reactivities of representative olefin types with the reagent. Accordingly, a number of studies were made in which the olefin in question was added to an equimolar quantity of disiamylborane in diglyme at 0°, and the rate of disappearance of the olefin followed by gas chromatography.[5]

Terminal olefins, such as 1-hexene, 3-methyl-1-butene, 2-methyl-1-butene, and 3,3-dimethyl-1-butene, react rapidly, with more than 90 per cent utilization of the olefin within $\frac{1}{2}$ hour.

Disubstituted internal olefins react much more slowly and exhibit remarkable differences. Thus cyclohexene reacts at a considerably lower rate than cyclopentene. The latter reacts at a slightly greater rate than *cis*-2-hexene. Quite remarkably, *trans*-2-hexene reacts at a rate considerably slower than that of the *cis* isomer. A bulky group on the double bond, as in *trans*-4-methyl-2-pentene and *trans*-4,4-dimethyl-2-pentene, leads to further reductions in rate.

Trisubstituted olefins, such as 2-methyl-2-butene, 1-methylcyclopentene, and 1-methylcyclohexene, react very slowly, as does the tetrasubstituted olefin, 2,3-dimethyl-2-butene. The experimental results on the reactivities of these representative olefins are summarized in Table 13-1.

These results establish the order of reactivity of olefins toward disiamylborane to be: 1-hexene $\geq$ 3-methyl-1-butene $>$ 2-methyl-1-butene $>$ 3,3-dimethyl-1-butene $>$ cyclopentene $\geq$ *cis*-2-hexene $>$ *trans*-2-hexene $>$ *trans*-4-methyl-2-pentene $>$ cyclohexene $\geq$ 1-methylcyclopentene $>$ 2-methyl-2-butene $>$ 1-methylcyclohexene $\geq$ 2,3-dimethyl-2-butene.

Not only do the reactivities vary markedly with the structure, but the differences in reactivities are quite large. For example, as will be shown in the next section, there is a factor of 100 between the relative reactivities of cyclopentene and cyclohexene. These large differences make possible some remarkably selective hydroborations, which will be discussed following review of the kinetic studies.

[9] H. C. Brown and B. C. Subba Rao, *J. Org. Chem.*, **22**, 1136 (1957).

Table 13-1

Reaction of Representative Olefins with Disiamylborane
in Diglyme at 0°

Olefin[a]	Olefin reacted, %				
	0.5 hr	1 hr	2 hr	4 hr	24 hr
1-Hexene	100				
3-Methyl-1-butene	99				
2-Methyl-1-butene	98				
3,3-Dimethyl-1-butene	93	98			
Cyclopentene	76	85	90		
cis-2-Hexene	62	79	93		
trans-2-Hexene	39	50	64	84	97
trans-4-Methyl-2-pentene	21	32	46	64	91
Cyclohexene	15	25	31	38	
1-Methylcyclopentene	12	19	25	34	
trans-4,4-Dimethyl-2-pentene	7	10	13	20	30
2-Methyl-2-butene	5	8	11		
1-Methylcyclohexene	4	6	7		
2,3-Dimethyl-2-butene	4	6	7		

[a] Olefin concentration was 0.5 M. An equivalent quantity of disiamylborane was used, partially in suspension.

It has recently been claimed that diethylborane is capable of achieving selective hydroborations in the same manner as disiamylborane.[10] The conclusion is therefore drawn that the large steric requirements of the siamyl group do not play any role in its selective reactions. Unfortunately, no experimental data supporting this position have been reported, so that it is not possible to arrive at an independent judgment of the validity of the conclusion.

Kinetics

Early attempts to follow the rate of reaction of diborane with representative olefins were unsuccessful.[11] The reaction was far too fast to follow in the ether solvents utilized for the hydroboration reaction. The observation that disiamylborane reacts with certain olefins at a reasonable rate prompted a study of the kinetics of the reaction in tetrahydrofuran as solvent, using cyclopentene as a representative olefin. It proved possible

[10] R. Köster and G. Griaznov, *Angew. Chem.*, **73**, 171 (1961).
[11] H. C. Brown and L. Case, unpublished studies.

to overcome the experimental difficulties and to obtain satisfactory, repro-ducible kinetic data.[12]

Disiamylborane exists in tetrahydrofuran solution as the dimer (Chap-ter 12).   The structure is presumably related to tetramethyldiborane with a double hydrogen bridge.[13]

$$\text{R} \diagdown \underset{\text{B}}{\diagup} \underset{\text{H}}{\diagdown} \underset{\text{B}}{\diagup} \diagdown \text{R}$$

The simplest possible mechanism for the reaction of this molecule with olefin would appear to involve a dissociation into a small equilibrium concentration of monomer, followed by a *cis* addition (Chapter 8) of the R_2BH monomer to the double bond.

$$R_2B \underset{H}{\overset{H}{\diamondsuit}} BR_2 \rightleftharpoons 2R_2BH$$

This mechanism requires three-halves-order kinetics.   However, the data do not obey this kinetic expression (Figure 13-1), so that this path appears to be ruled out for the present reaction.

The alternative mechanism in which the dimeric molecule reacts si-multaneously with two cyclopentene molecules requires third order ki-netics.

---

[12] H. C. Brown and A. W. Moerikofer, *J. Am. Chem. Soc.*, **83**, 3417 (1961).

[13] K. Hedberg, M. E. Jones, and V. Schomaker, Second International Congress of Crystallography, Stockholm, 1951, as reported in L. E. Sutton (ed.), "Tables of Interatomic Distances and Configuration in Molecules and Ions," The Chemical Society, London, 1958, p. M177.

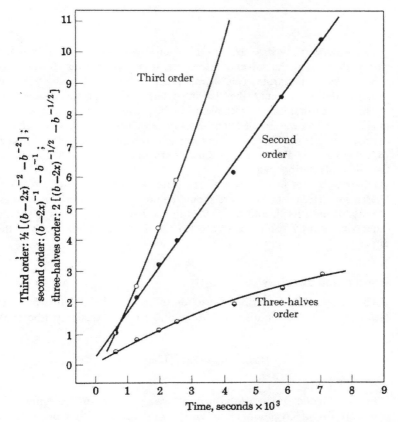

This path is also eliminated by the kinetics (Figure 13-1).

Figure 13-1 Kinetic analysis of the data for the reaction of disiamylborane dimer with cyclopentene in tetrahydrofuran at 0°.

The second-order kinetics (Figure 13-1) appear to require a reaction of the dimer with 1 cyclopentene molecule, presumably resulting in the formation of the product and 1 molecule of monomer. The latter presumably reacts with a second molecule of olefin, or dimerizes, in a rapid second step.

$$R_2BC_5H_{11} + R_2BH$$

This kinetic study was carried out in tetrahydrofuran, and no attempt was made to vary the solvent. However, it is probable that tetrahydrofuran plays an important role in this reaction by coordinating with the disiamylborane monomer, the "leaving group" in this particular reaction.

The reaction of diborane with olefins in the gas phase and in hydrocarbon media is very slow. Small quantities of various ethers greatly accelerate the reaction. It may be that this powerful catalytic effect of ethers is likewise due to their ability to coordinate with the borane leaving group in the initial alkylation stage.

Diborane exists in the monomeric state in tetrahydrofuran solution. Under these circumstances there should be no "leaving group" other than tetrahydrofuran itself, and the details of the hydroboration stage should differ significantly from these cases in which the reaction involves the dimer.

Structure and Reactivity

The kinetic data establish the reaction to be second order–first order in olefin and first order in disiamylborane dimer (sym-tetrasiamyldiborane).

$$\text{Rate} = k_2[\text{olefin}][(\text{Sia}_2BH)_2]$$

With the kinetic order established, it was possible to determine rate constants for the reaction of disiamylborane dimer with representative

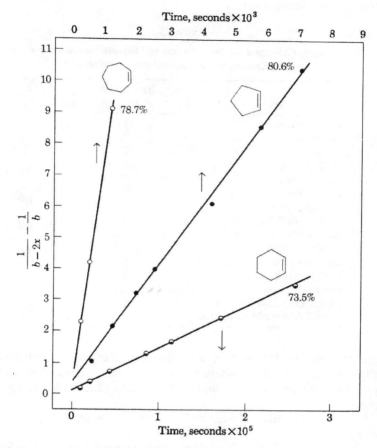

Figure 13-2 Second-order plots for the reaction of dimeric disiamylborane with cyclopentene, cyclohexene, and cycloheptene.

olefins (Figure 13-2), and to compare quantitatively the effect of structure upon the reactivity of various olefins. The rate constants established in this way are summarized in Table 13-2.

These results reveal a major change in reactivity in proceeding from cyclopentene to cyclohexene — a decrease of 100-fold. There then occurs an increase by a factor of over 500 with the seven-membered ring. From the rough data in Table 13-1, 1-methylcyclopentene possesses a reactivity almost equal to that of cyclohexene. It follows that the introduction of the methyl substituent reduces the rate by a factor of approximately 100. 1-Methylcyclohexene is exceedingly inert (Table 13-1). It may be assumed

Table 13-2

Rate Constants for the Reaction of Disiamylborane Dimer
with Olefins in Tetrahydrofuran Solution at 0°

Olefin	Rate constant $k_2 \times 10^4$, liter mole^{-1} sec^{-1}	$\dfrac{k_{cis}}{k_{trans}}$
Cyclopentene	14	
Cyclohexene	0.13	
Cycloheptene	72	
cis-2-Butene	23	6.0
trans-2-Butene	3.9	
cis-2-Pentene	21	7.0
trans-2-Pentene	3.0	
cis-4-Methyl-2-pentene	4.8	4.0
trans-4-Methyl-2-pentene	1.2	
cis-4,4-Dimethyl-2-pentene	0.78	7.8
trans-4,4-Dimethyl-2-pentene	0.10	
cis-3-Hexene	21	9.5
trans-3-Hexene	2.1	
cis-2,5-Dimethyl-3-hexene	1.1	2.5
trans-2,5-Dimethyl-3-hexene	0.42	

that the methyl substituent has a similar effect there, and possibly also in aliphatic systems.   The data reveal also that the difference in reactivity previously observed for cis- and trans-2-hexene (Table 13-1) is a general phenomenon.

The physical basis for these major differences in reactivity remains to be considered.

The addition of diethylaluminum hydride to the cycloalkenes, from cyclobutene through cycloundecene, has been examined.[14]   The reaction proceeds at a much slower rate than the addition of disiamylborane. Moreover, the variation of reactivity with ring size is considerably less, as indicated by the observed half-lives in minutes at 78°: cyclopentene, 200; cyclohexene, 1880; cycloheptene, 110.   However, the data reveal the same relative reactivity, cyclopentene > cyclohexene < cycloheptene, established in the hydroboration study.

Such a change in reactivity of the cyclic olefins with ring size appears to be general for reactions involving a simple cis addition to the double bond.   Thus the following half-lives (in minutes) have been observed for

[14] K. Ziegler, H. G. Gellert, H. Martin, K. Nagel, and J. Schneider, *Ann.*, **589**, 99 (1954).

the reaction of hexachlorocyclopentadiene with these olefins at 78°: cyclopentene, 80; cyclohexene, 1760, cyclopheptene, 61.[15]

The decrease of 100-fold observed in the reaction of disiamylborane with cyclopentene and cyclohexene might at first sight be attributed to the increased steric requirements of the relative nonplanar cyclohexene molecule as compared with the more planar cyclopentene system. However, the marked increase observed with cycloheptene argues against this factor as being the dominant one in these large variations in reactivity with ring size.

It is generally recognized that the cyclopentane and cycloheptane ring systems are strongly strained compared to cyclohexane itself.[16] Calculations based on the heats of hydrogenation of the corresponding cycloalkenes show that both cyclopentene and cycloheptene are strained as compared to cyclohexene.[17] Consequently, it appears that the increased reactivities of the five- and seven-membered ring olefins as compared to cyclohexene must be the result of the greater strain to which the double bonds are subjected in the cyclopentene and cycloheptene molecules.[18] This conclusion is supported by the observation that the more strained *trans* isomers of cyclononene and cyclodecene are far more reactive than the *cis* isomers toward diethylaluminum hydride.[18]

For the above interpretations to be acceptable, the transition stage must be very close to the olefin-disiamylborane structure. Cyclopentane and cycloheptane are considerably less stable thermodynamically than cyclohexane. If the transition states resembled products closely, the markedly lower stability of cyclopentyl- and cycloheptyldisiamylborane would have been expected to favor the reaction of the reagent with cyclohexene, leading to the more stable derivative, cyclohexyldisiamylborane.

The difference in the reactivities of *cis*- and *trans*-2-butene is likewise interpretable in terms of the greater strain present in the *cis* olefin.[19] Moreover, the phenomenon appears to be general for open-chain *cis-trans* olefins.

The reaction also appears to be sensitive to the steric requirements of

[15] K. Ziegler and H. Froitzheim-Kühlhorn, *Ann.*, **589**, 157 (1954).

[16] W. G. Dauben and K. S. Pitzer, Conformational Analysis, in M. S. Newman (ed.), "Steric Effects in Organic Chemistry," Chap. 1, John Wiley & Sons, Inc., New York, 1956.

[17] R. B. Turner and W. R. Meador, *J. Am. Chem. Soc.*, **79**, 4133 (1957).

[18] K. Ziegler, H. Sauer, L. Bruns, H. Froitzheim-Kühlhorn, and J. Schneider, *Ann.*, **529**, 122 (1954).

[19] G. B. Kistiakowsky, J. B. Ruhoff, H. A. Smith, and W. E. Vaughan, *J. Am. Chem. Soc.*, **57**, 876 (1935).

the olefin. As one of the methyl groups in *cis*-2-butene is varied from methyl to ethyl, to isopropyl, to *t*-butyl, the rate of reaction decreases, slowly at first but more rapidly with the more bulky substitutent ($k_2 \times 10^4$ liter mole^{-1} sec^{-1}): Me, 23; Et, 21; *i*-Pr, 4.8; *t*-Bu, 0.78. A similar effect is observed in the *trans* derivatives: Me, 3.9; Et, 3.0; *i*-Pr, 1.2; *t*-Bu, 0.10. Simultaneous variation in the structure of both groups exerts cumulative effects: *cis*-Me$_2$, 23; *cis*-Et$_2$, 21; *cis*, *i*-Pr$_2$, 1.1; *trans*-Me$_2$, 3.9; *trans*-, Et$_2$, 2.2; *trans-i*-Pr$_2$, 0.42.

Disiamylborane is a reagent of large steric requirements. This marked decrease in rate observed in both the *cis* and *trans* derivatives with an increase in the bulk of the alkyl substituents is attributed to the conflicting steric requirements of the two reactants in the transition state.

Consequently, it appears that both the strains to which the double bonds are subjected in the olefin and the steric requirements of the structure represent important factors in controlling the rate of reaction of disiamylborane with double bonds.

Selective Hydroborations

The major differences in the relative reactivity of various olefin structures should permit a wide variety of selective hydroboration reactions. Consequently, a number of competitive hydroborations were carried out to test the utility of the reagent for such selective reactions.[5]

Thus, treatment of an equimolar mixture of 1-pentene and 2-pentene with a controlled quantity of disiamylborane results in the preferential reaction of 1-pentene, leaving the 2-pentene in a purity of at least 99 per cent. Similarly, the reagent reacts preferentially with 2,4,4-trimethyl-1-pentene, producing 2,4,4-trimethyl-2-pentene containing no detectable trace of the isomeric olefin.

The reagent reacts preferentially with 1-hexene in a mixture with cyclohexene, yielding the cyclic olefin quantitatively free of the alkene. Similarly, the more reactive cyclopentene is readily hydroborated by the reagent in the presence of the relatively inert cyclohexene.

$$n\text{-}C_4H_9CH = CH_2$$

$$n\text{-}C_6H_{13}BSia_2$$

Finally, it has been observed that the reagent can distinguish between simple terminal olefins and those containing a methyl branch in the 2 position.

The remarkable difference in the relative reactivity of *cis* and *trans*-2-hexene is of special interest. It suggests the possibility that disiamylborane might provide a means of removing the *trans* isomer from an isomeric mixture to produce the pure *cis* compound. Actually, a commercial mixture of 2-pentene, analyzing 18 per cent *cis*- and 82 per cent *trans*-2-pentene, was converted by a single treatment into *trans*-2-pentene of 98 per cent purity.

$$sec\text{-}C_5H_{11}BSia_2$$

The observation that the rate of reaction of disiamylborane with olefins is very sensitive to the structure of the olefins makes possible the selective hydroboration of a more reactive olefin in the presence of a less reactive. However, the synthetic chemist is more frequently faced with the problem of two or more reactive sites within a single molecule. Fortunately, it has proved possible to utilize the data on the effect of structure on reactivity from simple olefins to predict the selective point of attack in molecules containing two or more reactive centers. For example, both 4-vinylcyclohexene and *d*-limonene undergo selective hydroboration at the predicted center.

A more detailed discussion of the utility of disiamylborane for the selective hydroboration of dienes and acetylenes will be presented in Chapters 15 and 16.

14 | Asymmetric Hydroboration with Diisopinocampheylborane

α-Pinene is available from natural sources in optically active form. It is the chief constituent of turpentine and can be isolated in both laevo- and dextrorotatory forms from turpentines obtained from different geographical locations.

As previously discussed, hydroboration of α-pinene involves a *cis* addition from the less hindered side[1,2] (Chapter 8). The reaction proceeds to the dialkylborane stage in diglyme, but appears to require excess α-pinene to achieve this stage in tetrahydrofuran, a solvent in which the product is more soluble (Chapter 11).

The stereochemistry of the addition is confirmed by oxidation of the intermediate dialkylborane into isopinocampheol. Thus, α-pinene ($[\alpha]_D^{20}$ +48°; lit.[3] $[\alpha]_D$ +51.1°) yields isopinocampheol ($[\alpha]_D^{20}$ −32.4°, m.p. 54–56°; lit.[4] $[\alpha]_D^{20}$ −32°, m.p. 57°). Consequently, the intermediate dialkylborane must have the structure described by the term, diisopinocampheylborane, existing in the dimeric form (Chapter 12).

[1] H. C. Brown and G. Zweifel, *J. Am. Chem. Soc.*, **81**, 247 (1959).
[2] H. C. Brown and G. Zweifel, *J. Am. Chem. Soc.*, **83**, 2544 (1961).
[3] F. H. Thurber and R. C. Thielke, *J. Am. Chem. Soc.*, **53**, 1030 (1931).
[4] H. Schmidt, *Ber.*, **77**, 544 (1944).

The observed rotation of diisopinocampheylborane dimer in tetrahydrofuran (with excess α-pinene to repress dissociation) is $[\alpha]_D^{20}$, $-37.1°$, with a value for the molecular optical rotation, $[M]_D^{20}$, of $-212.7°$.[5] Treatment with methanol converts diisopinocampheylborane into the methyl ester of diisopinocampheylborinic acid, with $[\alpha]_D^{20}$ $-53.6°$, $[M]_D^{20}$ $-169.3°$.[5]

This ready availability of an optically active dialkylborane suggested a study of the hydroboration of suitable olefinic derivatives in order to examine the possibility of achieving an asymmetric synthesis of an optically active organoborane, as well as of the alcohols derivable therefrom by oxidation.[6]

The results proved astounding. cis-2-Butene undergoes ready hydroboration by the reagent at 0° to form the organoborane, diisopinocampheyl-sec-butylborane. Oxidation produces a 90 per cent yield of 2-butanol. The observed rotation, $[\alpha]_D^{20}$ $-11.8°$, indicates an optical purity of 87 per cent (lit.[7] $[\alpha]_D^{20}$ $-13.51°$).

The use of laevorotatory α-pinene ($[\alpha]_D^{20}$ $-47.9°$) yields the dextrorotatory 2-butanol, $[\alpha]_D^{20}$ $+11.6°$.[8]

The treatment of cis-3-hexene with diisopinocampheylborane followed by oxidation yields 3-hexanol in a yield of 81 per cent. The observed rotation, $[\alpha]_D^{20}$ $-6.5°$ (lit.[9] $[\alpha]_D^{20}$ -7.13) indicates an optical purity of 91 per cent.

[5] H. C. Brown and G. J. Klender, J. Inorg. Chem., in press.

[6] H. C. Brown and G. Zweifel, J. Am. Chem. Soc., 83, 486 (1961).

[7] P. J. Leroux and H. J. Lucas, J. Am. Chem. Soc., 73, 41 (1951).

[8] H. C. Brown, G. Zweifel, and N. R. Ayyanger, manuscript in preparation.

[9] J. Kenyon and R. Poplett, J. Chem. Soc. (London), 1945, 273.

Application of the procedure to norbornene produces *exo*-norborneol.

The observed rotation, $[\alpha]_D^{20}$ -2.0; acetate, α_D^{20} $+7.9°$, indicates an optical purity of 83 per cent if the estimate is based on the value of Winstein and Trifan[10] ($[\alpha]_D^{20}$ $-2.41°$; acetate, α_D^{20} $+10.39°$) or an optical purity of 66 to 70 per cent from the conclusions of Berson and Suzuki.[11]

Finally, it should be mentioned that the procedure has been applied by Walborsky and Young to bicyclo [2:2:2] octene.[12] They realized bicyclo [2:2:2] octanol with an optical rotation, $[\alpha]_D^{25}$ $+6.7°$ as compared to the literature value of $[\alpha]_D^{25}$ $+7.45°$.[13]

The reagent reacts only very sluggishly with *trans* olefins or trisubstituted olefins, so that it does not appear to be as favorable for the conversion of these olefins into optically active derivatives. However, α-pinene is but one of a large number of available optically active terpenes. Presumably, it will be possible to tailormake many additional reagents more favorable for olefins of specific structural types.

The results clearly demonstrate that a boron atom at the asymmetric center is capable of maintaining asymmetry without significant racemization over periods of several hours. The ease with which organoboranes may be converted into other derivatives with retention of configuration

[10] S. Winstein and D. Trifan, *J. Am. Chem. Soc.*, **74**, 1154 (1952).

[11] J. A. Berson and S. Suzuki, *J. Am. Chem. Soc.*, **81**, 4088 (1959).

[12] H. M. Walborsky and A. E. Young, private communication.

[13] H. M. Walborsky, M. E. Baum, and A. A. Youssef, *J. Am. Chem. Soc.*, **83**, 988 (1961).

$$
\begin{array}{c}
\mathsf{H} \\
| \\
\mathsf{R-\overset{*}{C}-B} \!\!< \\
| \\
\mathsf{R'}
\end{array}
$$

and the unusually high optical purities achieved should make this approach to optically active derivatives a most valuable one for the chemist interested in synthesis.

Finally, it should be pointed out that the reagent prepared from dextrorotatory α-pinene yields (−) 2-butanol, (−) 3-hexanol, and (−) *exo*-norborneol. The absolute configurations of all these alcohols have been established [14],[15] and they can be correlated with the structure of the reagent. .Consequently, it appears that the reagent should not only be valuable as a practical tool for the synthesis of specific optically active derivatives but that it should also be very helpful for the more theoretical objective of establishing the absolute configurations of the products.

[14] J. A. Mills and W. Klyne, The Correlation of Configuration, in W. Klyne (ed.), "Progress in Stereochemistry," Vol. I, Chap. 5, p. 206, Butterworths Scientific Publications, London, 1954.

[15] J. A. Berson, J. S. Walia, A. Remanick, S. Suzuki, P. Reynolds-Warnhoff, and D. Willner, *J. Am. Chem. Soc.*, **83**, 3986, (1961).

15 | Hydroboration of Dienes

The hydroboration of olefins proceeds simply and quantitatively to the organoborane[1] (Chapter 6). However, it is evident that the extension of the reaction to dienes must involve problems not encountered with the simple olefins. First, the reaction of the difunctional diene with the polyfunctional diborane molecule would be expected to result in the formation of polymeric products, which might not exhibit the usual behavior of organoboranes. Second, conjugated dienes are less reactive toward simple addition reactions than related olefins. Consequently, the controlled monohydroboration of such dienes would appear to involve difficulties. However, a simple conversion of dienes into the corresponding diols or into the related unsaturated mono-ols would be a highly useful synthetic procedure. It appeared desirable, therefore, to explore the dihydroboration of representative dienes, as a possible synthetic route to diols, and a study of the monohydroboration of these dienes, as a possible synthesis of the corresponding unsaturated alcohols.[2-4]

Dihydroboration of 1,3-Butadiene

Diborane reacts vigorously with 1,3-butadiene in tetrahydrofuran solution, with a rapid utilization of the available hydride. Oxidation with alkaline hydrogen peroxide occurs normally, with a 75 to 80 per cent yield of butanediol isolated. Gas-chromatographic analysis of the product reveals a 24:76 distribution of the 1,3- and 1,4-diols. On the other hand,

[1] H. C. Brown and B. C. Subba Rao, *J. Am. Chem. Soc.*, **81**, 6428 (1959).
[2] H. C. Brown and G. Zweifel, *J. Am. Chem. Soc.*, **81**, 5832 (1959).
[3] G. Zweifel, K. Nagase, and H. C. Brown, *J. Am. Chem. Soc.*, **84**, 183 (1962).
[4] G. Zweifel, K. Nagase, and H. C. Brown, *J. Am. Chem. Soc.*, **84**, 190 (1962).

the reverse procedure, addition of the 1,3-butadiene to a solution of diborane in tetrahydrofuran, results in a hydroboration reaction which is relatively slow in the latter stages, with both free 1,3-butadiene and active "hydride" present. Oxidation of this product provides the 1,3- and 1,4-diols in a 35:65 ratio.

It was of interest to explore the nature of the organoborane from 1,3-butadiene and borane reacting in a ratio of 1:1. It appeared possible that the reaction would proceed as a simple cyclization to give cyclic diborane derivatives.

$$
2 \begin{array}{c} HC{=}CH_2 \\ | \\ HC{=}CH_2 \end{array} + B_2H_6 \longrightarrow
$$

$$
\begin{array}{ccccccc}
H_2C{-}CH_2 & & H & & CH_2{-}CH_2 \\
| & B & & B & & | \\
H_2C{-}CH_2 & & H & & CH_2{-}CH_2
\end{array}
$$

or

$$
\begin{array}{c}
CH_2{-}CH_2 \\
H_2C \quad H \quad CH_2 \\
B \quad B \\
H_2C \quad H \quad CH_2 \\
CH_2{-}CH_2
\end{array}
$$

Accordingly, the product from the 1:1 reaction in tetrahydrofuran solution was treated with methanol and subjected to distillation at low pressures. Hydrogen was evolved quantitatively on addition of the methanol, but only a negligible fraction of the product proved to be volatile.

This result clearly points to the formation of a polymeric organoborane in the initial hydroboration. This conclusion was further tested by performing the reaction of 1,3-butadiene and borane in a 1:1 ratio in a high-vacuum apparatus and then determining the molecular weight of the product in tetrahydrofuran solution by the lowering in vapor pressure of the solvent (Chapter 12). The molecular weight indicated was 320 to 365, nearly five times the value for the formation of a simple monomer, $(CH_2)_4BH$.

The high yield of the 1,3-butanediol isomer, the failure to realize a volatile methanolysis product, and the high molecular weight of the initial hydroboration product all point to the conclusion that the initial reaction is not simple cyclization, but involves the formation of a relatively complex polymer. Presumably, the slow reaction observed in the latter stages, especially when the butadiene is added to diborane, is due to the difficulties experienced by the free butadiene molecules in approaching the residual boron-hydrogen bonds entangled within a three-dimensional netlike polymer.

These observations that the hydroboration of 1,3-butadiene yields a polymer as the initial product appear to be in conflict with the reports of Köster.[5,6] In these communications it is reported that the reaction of butadiene with diborane yields 1,1'-tetramethylene-bis-borolane, $B_2(C_4H_8)_3$.

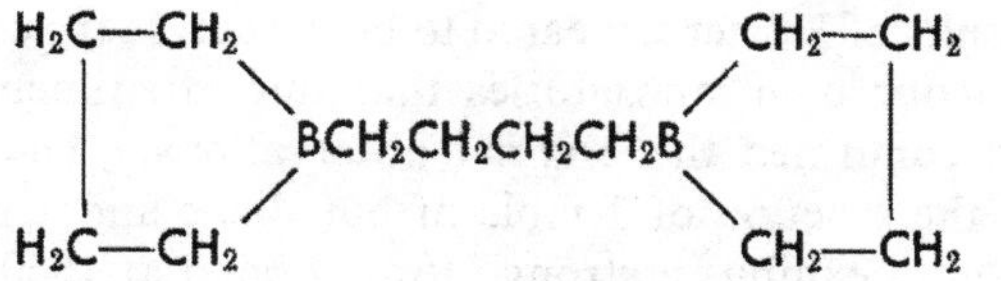

The formation of this compound is indicated as proceeding through the initial synthesis of borocyclopentane by the following scheme:

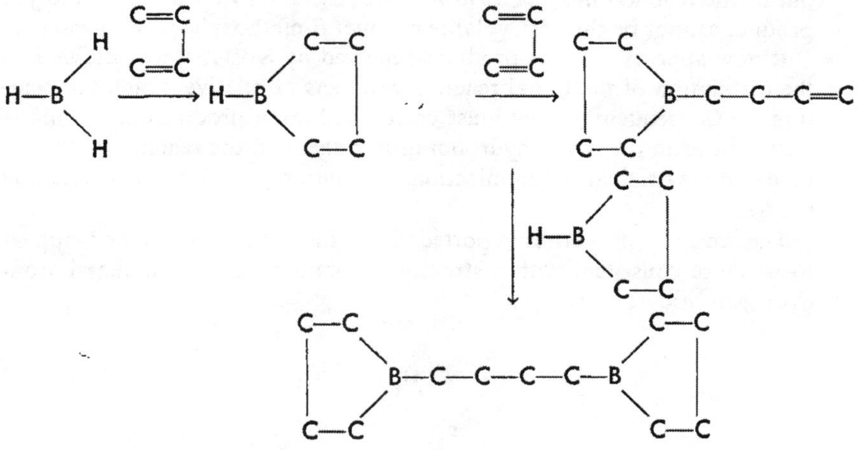

In the second communication[6] Köster reports the isolation of *bis*-boracyclopentane from the reaction of 1,3-butadiene and diborane at room temperature.

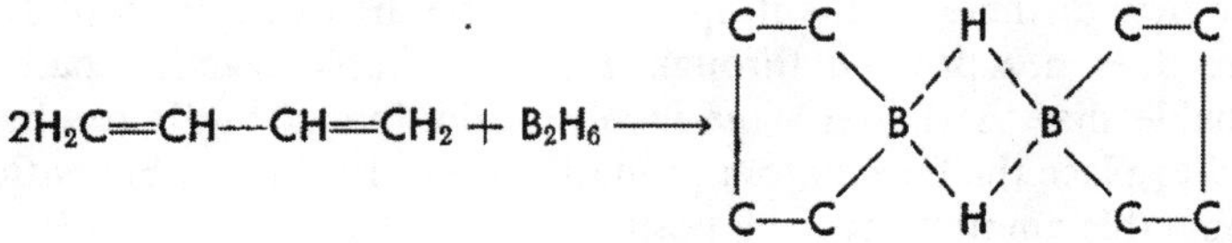

$$2H_2C=CH-CH=CH_2 + B_2H_6 \longrightarrow$$

[5] R. Köster, *Angew. Chem.*, **71**, 520 (1959).
[6] R. Köster, *Angew. Chem.*, **72**, 626 (1960).

The properties of this product are unusual: no reaction with olefins except at temperatures over 70°, no reaction with methanol under 100°; infrared absorption for the hydrogen bridge at 1612 cm^{-1}. In contrast,[7] tetrasi-amyldiborane reacts rapidly with terminal olefins at 0°, reacts rapidly with methanol at 0°, and exhibits the normal bridge absorption at 1555 cm^{-1} (Chapter 12).

These results of Köster appeared to be in such sharp contrast to the data obtained in our own laboratories that our experiments were repeated. The results confirmed the original observations. The reaction mixture formed by the reaction of 1 mole of butadiene and 1 mole of borane in tetrahydrofuran exhibits a strong infrared band at 1560 cm^{-1} and a band of medium intensity at 2500 cm^{-1}, assigned to the bridge hydrogens and a terminal boron-hydrogen structure. The product reacts rapidly with methanol at room temperature. The methanolysis product does not distill out of the reaction mixture at low pressures. Evidently the methanolysis product cannot be the very volatile product β-methoxyboracyclopentane.[6]

It now appears that the products reported by Köster were obtained by the distillation of the initial reaction products at relatively high temperatures.[8] Consequently, they must be redistribution products of the materials formed in the initial hydroboration stage, and the reaction path suggested in the original communications[5,6] cannot represent the true reaction course.

The unusual properties reported[6] for "bis-boracyclopentane" appear to be more consistent with a structure containing a transannular boron-hydrogen bridge,

than with the structure proposed by Köster.

In view of these results it appears that the hydroboration of 1,3-buta-diene does not proceed through a single, simple reaction path. It is probable that the initial stage involves addition of the boron-hydrogen bond to place the boron atom primarily at the 1 position, but with nearly comparable amounts at the 2 position.

[7] H. C. Brown and G. J. Klender, *J. Inorg. Chem.*, in press.
[8] R. Köster, private communication.

$$CH_2\!\!=\!\!CH\!\!-\!\!CH\!\!=\!\!CH_2 + BH_3 \longrightarrow CH_2\!\!=\!\!CH\!\!-\!\!CH_2\!\!-\!\!CH_2$$
$$\underset{\displaystyle BH_2}{|}$$

$$CH_2\!\!=\!\!CH\!\!-\!\!CH\!\!=\!\!CH_2 + BH_3 \longrightarrow CH_2\!\!=\!\!CH\!\!-\!\!CH\!\!-\!\!CH_3$$
$$\underset{\displaystyle BH_2}{|}$$

The unusually large amount of addition at the secondary position is probably the result of the same factors which operate in the hydroboration of styrene (Chapter 7).

The second stage presumably involves a second addition, favored by the fact that the now-isolated double bond is more reactive than the conjugated diene.

$$2CH_2\!\!=\!\!CHCH_2CH_2 \longrightarrow CH_2\!\!=\!\!CHCH_2CH_2$$

with BH_2 below, leading to

$$CH_2\!\!=\!\!CHCH_2CH_2$$
$$\diagdown BH$$
$$CH_2CH_2CH_2CH_2$$
$$\underset{\displaystyle BH_2}{|}$$

$$CH_2\!\!=\!\!CHCH_2CH_2 + CH_2\!\!=\!\!CHCHCH_3 \longrightarrow$$
with BH_2 under each reactant

$$CH_2\!\!=\!\!CHCH_2CH_2$$
$$\diagdown BH$$
$$CH_3CHCH_2CH_2$$
$$\underset{\displaystyle BH_2}{|}$$

Cyclization doubtless occurs in an appreciable fraction of the cases when the double bond and boron-hydrogen reaction centers are suitably situated.

With the boron atom possessing three hydrogen atoms capable of adding to the bifunctional butadiene molecule, this series of competing addition reactions would lead to a complex polymer with the observed properties. Oxidation would yield a mixture of 1,3- and 1,4-butanediols, as observed. Moreover, a change in the mode of addition would be expected to alter the relative importance of the different paths and bring about a change in the relative amounts of 1,3- and 1,4-diols formed.

Dihydroboration with Diborane

Based upon the results realized with 1,3-butadiene, a standard procedure was developed for the dihydroboration of dienes and applied to a number of representative dienes. In this procedure a solution of diborane in tetrahydrofuran was added at 20 to 25° to the diene in the same solvent. After 1 to 2 hours, the product was oxidized at 30 to 50° with alkaline hydrogen peroxide. Saturation of the aqueous phase with potassium carbonate transferred the glycol into the tetrahydrofuran phase. Typical results are reported in Table 15-1.

Table 15-1
Dihydroboration of Dienes with Diborane — Synthesis of Diols

Diene	Diene, mmoles	Borane, mmoles	G.C. analysis[a]	Product[b]
1,3-Butadiene	200	200	24% 1,3- 76% 1,4-	75% (1,3- and 1,4-)
2-Methyl-1,3-butadiene	150	100	13% 1,3- 87% 1,4-	68% (1,3- and 1,4-)
2,3-Dimethyl-1,3-butadiene	100	100	100% 1,4-	66% 1,4-
1,4-Pentadiene	150	100	62% 1,4- 38% 1,5-	41% 1,4-
1,5-Hexadiene	150	100	22% 1,5- 69% 1,6- 9% 2,5-	58% 1,6-
1,3-Cyclohexadiene	150	100	Two peaks	61%
Bicycloheptadiene	100	100	One peak	64%

[a] Generally of crude reaction product, before distillation.
[b] Isolated.

As was pointed out in Chapter 7, the presence of an alkyl group in the 2 position of a terminal olefin markedly favors placement of the boron at the terminal carbon atom. This effect is reflected in the results with 2-methyl-1,3-butadiene and 2,3-dimethyl-1,3-butadiene. Whereas 1,3-butadiene provides approximately a 70:30 distribution of the two isomers, the product from 2-methyl-1,3-butadiene contains 87 per cent of 2-methyl-1,4-butanediol, while 2,3-dimethyl-1,3-butadiene yields essentially pure 2,3-dimethyl-1,4-butanediol.

The dihydroboration of 1,4-pentadiene yields a 62:38 distribution of 1,4- and 1,5-pentanediol. The hydroboration of a simple terminal olefin places 6 per cent of the boron on the secondary position, 94 per cent on the primary. The predominant formation of 1,4-pentanediol with its secondary alcohol group can be understood only on the basis of an intramolecular addition of the second boron-hydrogen bond with the formation of a five-membered cyclic organoborane. On the basis of the usual four-center transition state (Chapter 8), such an addition would, on stereochemical grounds, be much more favorable than in the case of 1,3-butadiene.

$$CH_2\!\!=\!\!CHCH_2CH\!\!=\!\!CH_2 \xrightarrow{\ BH_3\ } CH_2\!\!=\!\!CHCH_2CH_2CH_2$$

Dihydroboration of 1,5-hexadiene, followed by oxidation, yields a mixture of diols, the main product being the 1,6-hexanediol (69 per cent). Simple dihydroboration of 1,5-hexadiene would have been expected to yield no more than 6 per cent of the 1,5-hexanediol. The observed formation of 22 per cent of this product argues for the incursion of a cyclic mechanism here also. However, such a cyclic mechanism would involve a six-membered ring, and it is apparently considerably less favorable than the five-membered ring postulated to be one of the intermediates in the hydroboration of 1,4-pentadiene.

The dihydroboration of 1,3-cyclohexadiene and bicycloheptadiene results in the formation of a polymeric organoborane which forms a gel. However, the products readily undergo oxidation to diols, whose precise structures have not been established. The dihydroboration of cyclopentadiene provides cis-1,3-cyclopentanediol.[9]

[9] K. A. Saegebarth, *J. Org. Chem.*, **25**, 2212 (1960).

Dihydroboration with Disiamylborane

Disiamylborane would appear to have two major advantages for the dihydroboration of dienes.[4] It is a monofunctional reagent which should avoid the formation of the complex polymers observed with diborane. It exhibits a greater ability to place the boron atom at the terminal carbon of the double bond (Chapter 13).

Actually, the dihydroboration of 1,3-butadiene with disiamylborane yields a 10:90 distribution of the 1,3- and 1,4-butanediol, as compared with the 30:70 distribution realized with diborane. Similarly, diborane converts 1,4-pentadiene predominantly (62 per cent) into the 1,4-diol. Disiamylborane provides 85 per cent of the 1,5-pentanediol. Finally, 1,5-hexadiene, which yields with diborane 69 per cent 1,6-hexanediol and 31 per cent isomeric diols, provides 93 per cent 1,6- and only 7 per cent isomeric derivatives.

These results clearly demonstrate the advantages of disiamylborane for the hydroboration of simple terminal dienes. Of course, in cases where the diene contains alkyl substitutents in the alpha position, such as 2,3-dimethyl-1,3-butadiene, there is a much more powerful influence directing the boron atom to the terminal position, and hydroboration with diborane alone will provide excellent yields of an isomerically pure product.

Dihydroboration with Trimethylamine-t-butylborane

Hawthorne has recently reported that 1,3-butadiene, 1,4-pentadiene and 2-methyl-1,3-butadiene, react at 60° with trimethylamine-t-butylborane.[10] Distillation of the reaction products yields volatile organoboranes, identified as the 1-t-butyl-1-borocycloalkanes. Yields of 40 to 60 per cent were realized.

[10] M. F. Hawthorne, *J. Am. Chem. Soc.*, **83**, 2541 (1961).

Unfortunately, neither the initial hydroboration products nor the distilled 1-*t*-butyl-1-borocycloalkanes were oxidized, so it is not possible to state how fully this reagent reacts to place the boron atom at the terminal position. If the reaction proceeds quantitatively as indicated, the reaction would be a valuable one for the hydroboration of dienes.

Isomerization of Organoboranes from Dienes

As discussed earlier in this chapter, the initial reaction of diborane with dienes in the stoichiometric ratio (three dienes per B_2H_6) appears to lead to the formation of a complex polymer. Distillation at elevated temperatures appears to result in a redistribution of the groups, with the formation of bis-(1-boracycloalkyl) alkanes.[5]

$$3n C_4H_6 + n B_2H_6 \longrightarrow (C_4H_8)_{3n}B_{2n}$$

$$\Big\downarrow \Delta$$

Saegebarth has explored the behavior of these products under isomerization conditions (Chapter 9) with interesting results.[11] For example, the hydroboration product from 1,3-pentadiene was distilled to yield the corresponding bis-(1-boracycloalkyl)alkane. Oxidation with alkaline hydrogen peroxide provided 10 per cent of 1,3- and 90 per cent of 1,4-pentanediol. However, after isomerizing at 160 to 175° for 6 hours, the oxidation yielded essentially 100 per cent 1,5-pentanediol.

Similarly, the isomerized organoborane from 1,5-hexadiene gave 70 per cent 1,5- and 30 per cent 1,6-hexanediol. The same mixture was obtained in isomerizing the organoborane from 1,4-hexadiene.

These results suggest that a six-membered boron heterocycle is more stable thermodynamically than either a five- or seven-membered ring.

[11] K. A. Saegebarth, *J. Am. Chem. Soc.*, **82**, 2081 (1960).

Monohydroboration with Diborane

In order to examine the possibility of achieving the monohydroboration of dienes with diborane,[3] a quantitative study was made of the reaction of equivalent quantities of borane and representative dienes. In these experiments, 25 mmoles of the diene was added to tetrahydrofuran containing 8.4 mmoles of borane. At the end of 1 and 2 hours, samples were removed and analyzed for residual diene. The yield of the monohydroboration product was estimated from the amount of residual diene. It is evident that complete monohydroboration (100 per cent yield) would involve the complete utilization of diene; in the case of complete dihydroboration (0 per cent yield of monohydroboration) there would remain 50 per cent of the original diene. The results are summarized in Table 15-2.

Table 15-2
Stoichiometry of the Reaction of Dienes with Diborane in Amounts Corresponding to Monohydroboration[a]

Diene	Residual diene, %		Monohydroboration, %
	1 hr	2 hr	
1,3-Butadiene	50	48	4
2-Methyl-1,3-butadiene	51	48	4
1,3-Pentadiene	43	44	12
1,4-Pentadiene	41	45	10
1,5-Hexadiene	33	32	35
1,3-Cyclohexadiene		24	51
Bicycloheptadiene	33	33	34
1,5-Cyclooctadiene	51	45	10

[a] 25 mmoles of diene added to 8.4 mmoles of borane.

These results indicate that the preparation of unsaturated alcohols via hydroboration with diborane would not be practical for conjugated dienes. Such dienes undergo dihydroboration preferentially. This result appears to arise from the reduced activity of the conjugated diene system toward hydroboration. For example, the competitive hydroboration of a mixture of 1,3-butadiene and 1-hexene results in the preferential reaction of the olefin. Consequently, once monohydroboration of the diene has occurred, there is formed an unsaturated organoborane with an isolated double bond. This product is more reactive than the diene and undergoes further hydroboration preferentially.

$$CH_2\!\!=\!\!CHCH\!\!=\!\!CH_2 \xrightarrow[\substack{\text{relatively}\\\text{slow}}]{HB} CH_2\!\!=\!\!CHCH_2CH_2\!\!-\!\!B\!\!<$$

$$\text{fast} \downarrow HB$$

$$C_4H_8\big(B\!\!<\big)_2$$

The results are more favorable for the unconjugated dienes. For example, 1,5-hexadiene yields very nearly the statistical distribution of 50 per cent mono- and 25% dihydroboration. In the case of 1,4-pentadiene the results are low. Either there is some interaction between the double bonds, or the cyclic mechanism discussed earlier intervenes.

The results appear favorable for realizing monohydroboration in the cyclic dienes, whether the diene is conjugated or not. Even in these cases there is evident a competition between the two hydroboration stages, so that it would appear desirable to utilize an excess of the diene to favor the monohydroboration stage.

In this way 1,5-hexadiene was converted into 5-hexene-1-ol in 26 per cent yield.

$$\begin{array}{l} CH_2CH\!\!=\!\!CH_2 \\ | \\ CH_2CH\!\!=\!\!CH_2 \end{array} \xrightarrow{HB} \xrightarrow{[O]} \begin{array}{l} CH_2CH\!\!=\!\!CH_2 \\ | \\ CH_2CH_2CH_2OH \end{array}$$

Winstein and his co-workers have realized the isolation of 3-cyclopentene-1-ol in 30 per cent yield by the monohydroboration of cyclopentadiene.[12]

[12] E. L. Allred, J. Sonnenberg, and S. Winstein, *J. Org. Chem.*, **25**, 26 (1960).

1,3-Cyclohexadiene provides a 34 per cent yield of cyclohexenol. However, gas-chromatographic analysis indicated the formation of two alcohols: 60 to 65 per cent of 2-cyclohexene-1-ol and 35 to 40 per cent of 3-cyclohexene-1-ol. The reason for the difference in the results from those realized with cyclopentadiene is not apparent. Possibly it arises from the effect of the puckered methylene group in 1,3-cyclohexadiene in directing the boron atom away from the adjacent carbon atom of the diene system. This effect would be found to be much smaller in the planar cyclopentadiene molecule.

Finally, the monohydroboration of bicycloheptadiene provides a convenient synthesis of *exo*-dehydronorborneol in satisfactory yield (63 per cent crude, 45 per cent isolated).

However, the same procedure applied to 1,5-cycloöctadiene results in preferential dihydroboration.

Monohydroboration with Disiamylborane

The use of disiamylborane should offer no significant advantage over diborane in the monohydroboration of acyclic dienes, such as 1,3-butadiene, where the two double bonds are identical. Just as in the diborane case, the monohydroboration stage produces a more-reactive unsaturated organoborane, which should readily react further. On the other hand, in cases where the two double bonds possess structures that differ markedly in their reactivity toward disiamylborane (Chapter 13), the latter would be expected to react preferentially at the more reactive of the two sites in the diene, converting the latter into an organoborane with a double bond that is relatively inert toward the reagent. In such cases, disiamylborane might be expected to lead to an improvement in yield of the monohydroboration derivative.

$$CH_2\!\!=\!\!CHCH\!\!=\!\!CH_2 \xrightarrow[\substack{\text{relatively}\\ \text{slow}}]{Sia_2BH} CH_2\!\!=\!\!CHCH_2CH_2BSia_2$$

$$\text{fast} \Big\downarrow Sia_2BH$$

$$Sia_2BCH_2CH_2CH_2CH_2BSia_2$$

$$CH_3CH\!\!=\!\!CHCH\!\!=\!\!CH_2 \xrightarrow[\substack{\text{relatively}\\ \text{slow}}]{Sia_2BH} CH_3CH\!\!=\!\!CHCH_2CH_2BSia_2$$

$$\text{slow} \Big\downarrow Sia_2BH$$

dihydroboration

Finally, in some cases, cyclization of the initially formed monoalkenylborane appears to be responsible for the poor yields realized with diborane. Here also the use of disiamylborane should improve the situation.

These predictions are confirmed by the stoichiometry studies summarized in Table 15-3. (The experiments were carried out in the same manner as described previously for the diene-diborane stoichiometry studies reported in Table 15-2).

It should be emphasized that the above yields were realized in systems in which the diene and disiamylborane were present in equimolar amounts. The use of an excess of the diene should reduce greatly the amount of material diverted to the dihydroboration product.

Study of the monohydroboration of dienes with diborane had indicated that the behavior of the cyclic dienes was quite different from that of the acyclic derivatives. Reasonable yields were realized for several of the cyclic monohydroboration products (Table 15-2). The use of disiamylborane provides excellent yields of the monohydroboration product with both 1,3- and 1,4-cyclohexadiene. However, just as in the case with diborane, 1,5-cyclooctadiene undergoes dihydroboration preferentially (Table 15-3).

Table 15-3

Stoichiometry of the Reaction of Representative Dienes with Disiamylborane at 0°[a]

Diene	Residual diene, %				Monohydroboration, %	
	0.5 hr	1 hr	2 hr	8 hr	Disiamylborane[b]	Diborane[c]
1,3-Butadiene		43	46		8	4
2-Methyl-1,3-butadiene	38	38			24	4
1,3-Pentadiene	16	11	12		76	12
1,4-Pentadiene		34	35		30	10
1,5-Hexadiene	25	24			52	35
2-Methyl-1,5-hexadiene	25	17	16		68	
1,3-Cyclohexadiene	62	49	6		88	51
1,4-Cyclohexadiene	56	50	40	20[d]	82	
1,5-Cycloöctadiene	55	47	46		6	10
Bicycloheptadiene	59	44	40	30	40	34
Cyclohexene[e]		82	77	61		
Cyclooctene[e]	6					

[a] The diene, 25 mmoles, was treated with 25 mmoles of disiamylborane at 0 to 5°.   [b] The percentage of monohydroboration is based on the last analysis for residual diene.   [c] Table 15-2.   [d] After 24 hr, 9%.   [e] 25 mmoles olefin and 25 mmoles disiamylborane.

The initial product of the hydroboration of 1,5-cycloöctadiene must be a cycloöctene derivative.

Cycloöctene itself is very reactive toward disiamylborane (Table 15-3). Nevertheless, the failure to realize any significant yield of the hydroboration product indicates that the diene is less reactive than the olefin. Such a decreased reactivity might arise as a result of transannular conjugation between the two double bonds, or it might be the result of lower strain in the cyclooctadiene arising from the lower number of carbon-hydrogen interactions.[13]

Köster reports that the treatment of 1,5-cyclooctadiene with diethylborane likewise results in a dihydroboration product.[14] This product undergoes an interesting transformation under the influence of heat in the presence of catalytic quantities of boron hydride bonds.

Very satisfactory yields have been realized on a preparative scale, with many of the dienes examined, utilizing 100 per cent excess of diene. Thus, 1,3-pentadiene provides a 74 per cent yield of 3-pentene-1-ol, and 2-methyl-1,5-hexadiene 71 per cent of 5-methyl-5-hexene-1-ol. In each case the disiamylborane reacts at the less hindered of the two positions.

$$CH_3CH{=}CHCH{=}CH_2 \xrightarrow{Sia_2BH} \xrightarrow{[O]} CH_3CH{=}CHCH_2CH_2OH$$

$$\underset{CH_2{=}CCH_2CH_2CH{=}CH_2}{\overset{CH_3}{|}} \xrightarrow{Sia_2BH} \xrightarrow{[O]} \underset{CH_2{=}CCH_2CH_2CH_2CH_2OH}{\overset{CH_3}{|}}$$

Similarly, 1,5-hexadiene provides a 64 per cent yield of 5-hexene-1-ol.

$$\underset{CH_2CH{=}CH_2}{\overset{CH_2CH{=}CH_2}{|}} \xrightarrow{Sia_2BH} \xrightarrow{[O]} \underset{CH_2CH_2CH_2OH}{\overset{CH_2CH{=}CH_2}{|}}$$

[13] W. G. Dauben and K. S. Pitzer, Conformational Analysis, in M. S. Newman (ed.), "Steric Effects in Organic Chemistry," Chap. 1, John Wiley & Sons, Inc., New York, 1956.

[14] R. Köster and G. Griaznov, *Angew. Chem.*, **73**, 171 (1961).

In the event the second position is relatively inert toward disiamylborane, excellent yields can be realized even without an excess of diene. Thus, 4-vinylcyclohexene and *d*-limonene are readily converted to the corresponding unsaturated alcohols in high yields.[15]

The monohydroboration of 1,4-cyclohexadiene provides 3-cyclohexene-1-ol in high yield, 90 per cent.

Similarly, 1,3-cyclohexadiene gives the corresponding cyclohexenol in very high yield. However, in contrast to the results with diborane, where the distribution is approximately 35 to 40 per cent of 3- and 60 to 65 per cent of 2-cyclohexene-1-ol, the product from disiamylborane was predominantly (90 to 93 per cent) 2-cyclohexene-1-ol.

[15] H. C. Brown and G. Zweifel, *J. Am. Chem. Soc.*, **83**, 1241 (1961).

The most reasonable explanation is based upon the steric influence of the methylene hydrogen atoms adjacent to the diene system, which directs the bulky reagent to the more remote 3 position.

The effect would be less with diborane, so that the attack on the two positions become more nearly equivalent.

In support of this tentative explanation is the somewhat higher reactivity of the conjugated diene, 1,3-cyclohexadiene, as compared with the unconjugated derivative, 1,4-cyclohexadiene, or with the simple olefin, cyclohexene (Table 15-3).

It may be noted that both 1,4-cyclohexadiene and cyclohexene possess such a methylene group which could hinder the addition of the bulky disiamylborane group to the adjacent carbon atom. On the other hand, the disiamylboryl group can avoid such steric interactions with the methylene group by adding to the 3 position of the 1,3-cyclohexadiene.

Finally, it should be mentioned that disiamylborane promises to be a very useful reagent in even more complex systems. Applied to myrcene, it achieves the selective hydroboration of the less substituted terminal double bond.[16]

[16] H. C. Brown and K. P. Singh, unpublished results.

Similarly, it achieves the monohydroboration of caryophyllene. However, in this case the very high reactivity of the *trans* double bond in the nine-membered ring causes it to participate in preference to the exocyclic structure.

Conclusion

It is evident that the possibilities in the application of hydroboration to dienes and polyenes are largely untouched. It will require an intensive effort to reach a full understanding of the reactions possible and of the means for their control. However, it is evident that such studies will doubtless bring forth many new synthetic possibilities. The hydroboration of dienes and polyenes must be considered to be still a largely virgin field awaiting exploration.

16 | Hydroboration of Acetylenes

Vinylboron compounds have only recently become available through the reaction of vinylsodium and dimethylboron bromide.[1] A mixture of dimethylvinylborane, methyldivinylborane, and trivinylborane is obtained.

$$CH_2{=}CHNa + (CH_3)_2BBr \longrightarrow (CH_3)_2BCH{=}CH_2 +$$
$$CH_3B(CH{=}CH_2)_2 +$$
$$B(CH{=}CH_2)_3$$

Although there appears to be some question as to the precise properties and characteristics of trivinylborane,[2,3] there is little doubt that such vinyl organoboranes have interesting possibilities in organic synthesis. It was evident that the hydroboration reaction would provide a very simple route to these derivatives, providing it proved possible to halt the reaction at the monohydroboration stage.

[1] T. D. Parsons, M. B. Silverman, and D. M. Ritter, *J. Am. Chem. Soc.*, **79**, 5091 (1957).

[2] A. V. Topchiev, A. M. Payshkin, and A. A. Prokorova, *Doklady Akad. Nauk SSSR*, **129**, 598 (1959).

[3] F. E. Brinckman and F. G. A. Stone, *J. Am. Chem. Soc.*, **82**, 6218 (1960).

A study of the hydroboration of representative acetylenes has established the practicality of this synthesis.[4,5] It also proved of interest to explore the chemistry of the dihydroboration product.[5]

Monohydroboration of Acetylenes

The possibility of achieving monohydroboration was explored by following the reaction of acetylenes with a stoichiometric amount of hydroborating agent. The procedure was similar to that described for the corresponding reaction of the dienes (Chapter 15).

3-Hexyne and 1-hexyne were selected as typical representatives of acetylenes with an internal and a terminal triple bond. Each of the acetylenes was added to the usual hydroboration mixture of sodium borohydride in diglyme and the hydroboration accomplished at 0° by adding boron trifluoride etherate to the reaction mixture. Sufficient reagent was utilized to achieve the monohydroboration of the acetylene.

After 2 hours the solutions were found free of hydride. Gas chromatographic analysis revealed 84 per cent utilization of 3-hexyne and 44 per cent of 1-hexyne. Consequently, 3-hexyne reacts predominantly to form the desired trivinylborane derivative, with approximately 16 per cent of dihydroboration. On the other hand, the dominant reaction of 1-hexyne is dihydroboration.

In the case of 3-hexyne, it appears that the acetylene is far more reactive than the initial reaction product. As a result, the reaction proceeds satisfactorily to the formation of the desired trivinylborane.

$$
\begin{array}{c}
C_2H_5 \\
| \\
C \\
3\;\;||| \quad + BH_3 \quad \longrightarrow \\
C \\
| \\
C_2H_5
\end{array}
\qquad
\begin{array}{c}
C_2H_5\!\!-\!\!C\!\!-\!\!H \\
|| \\
C_2H_5\!\!-\!\!C\!\!-\!\!)_3B
\end{array}
$$

$$
\text{relatively} \Big|\; HB \atop \text{slow} \Bigg\downarrow
$$

dihydroboration

On the other hand, in the case of 1-hexyne, the initially formed product is apparently more reactive toward the reagent than the residual acetylene. In this case the 1-hexyne undergoes dihydroboration preferentially.

[4] H. C. Brown and G. Zweifel, *J. Am. Chem. Soc.*, **81**, 1512 (1959).
[5] H. C. Brown and G. Zweifel, *J. Am. Chem. Soc.*, **83**, 3834 (1961).

$$n\text{-}C_4H_9\text{---}\underset{\underset{\displaystyle H}{|}}{\overset{\overset{\displaystyle n\text{-}C_4H_9}{|}}{\underset{\|}{\overset{\displaystyle C}{C}}}} + H\text{---}B\diagup \longrightarrow \begin{array}{c} n\text{-}C_4H_9\text{---}C\text{---}H \\ \| \\ H\text{---}C\text{---}B\diagup \end{array}$$

fast $\Big|$ HB

dihydroboration

Probably the monohydroboration of 1-hexyne and related terminal acetylenes could be achieved by utilizing a very large excess to repress the second stage. Fortunately, a more convenient solution was developed. The use of disiamylborane (Chapter 13) provided essentially quantitative conversion of both 1- and 3-hexyne (Table 16-1).

Table 16-1

Monohydroboration of 3-Hexyne and 1-Hexyne
with Sodium Borohydride–Boron Trifluoride
and Disiamylborane

	Acetylene, mmoles	Hydroborating agent	Hydride added, mmoles	Acetylene reacted, mmoles	Monohydroboration, %
3-Hexyne	25	NaBH$_4$, BF$_3$[a]	25	21	66
3-Hexyne	25	Sia$_2$BH[b]	25	25	100
1-Hexyne	25	NaBH$_4$, BF$_3$[a]	25	14	12
1-Hexyne	25	Sia$_2$BH[b]	25	25	100

[a] Internal hydroboration in diglyme.
[b] Hydroboration with disiamylborane in tetrahydrofuran.

It would appear that the large steric requirements of the disiamylborane structure is largely responsible for the clean monohydroboration achieved.

$$\underset{\underset{\displaystyle R}{|}}{\overset{\overset{\displaystyle R}{|}}{\underset{\|}{\overset{\displaystyle C}{C}}}} + Sia_2BH \longrightarrow \begin{array}{c} R\text{---}C\text{---}H \\ \| \\ R\text{---}C\text{---}BSia_2 \end{array}$$

$$\begin{array}{c} \text{R} \\ | \\ \text{C} \\ ||| \\ \text{C} \\ | \\ \text{H} \end{array} + \text{Sia}_2\text{BH} \longrightarrow \begin{array}{c} \text{R---C---H} \\ || \\ \text{H---C---BSia}_2 \end{array}$$

Indeed, it was observed that the reaction of the initial reaction product with a second equivalent of disiamylborane is negligibly slow.

Recently, Hawthorne has reported that trimethylamine-t-butylborane likewise achieves the monohydroboration of several terminal acetylenes (1-butyne, 1-pentyne, and 1-hexyne).[6]

$$2\text{RC}\equiv\text{CH} + (\text{CH}_3)_3\text{N}:\text{BH}_2\text{C}(\text{CH}_3)_3 \longrightarrow (\text{RCH}=\text{CH})_2\text{BC}(\text{CH}_3)_3 + (\text{CH}_3)_3\text{N}$$

These products were isolated by fractional distillation in yields of 65 to 70 per cent.

The application of hydroboration to convert acetylenes to vinyl organoboranes has now been applied to a modest number of derivatives: 1-butyne,[6] 1-pentyne,[6] 2-pentyne,[5] 1-hexyne,[5,6] 2-hexyne,[5] 3-hexyne,[5] 1-heptyne,[7] 1-octyne,[5] and cyclodecyne.[8] The hydroboration of di-t-butyl-acetylene has also been reported,[9] but it is not clear whether the authors achieved the monohydroboration or dihydroboration stage.

Protonolysis of the Vinyl Organoboranes

The protonolysis of the trialkylboranes with acetic acid proceeds readily for the first group[10] but requires elevated temperatures for the second and third groups.[11] However, the protonolysis of the unsaturated organoboranes produced via the hydroboration of acetylenes proceeds to completion readily with acetic acid at room temperature. By this procedure, 2-pentyne and 3-hexyne were converted to cis-2-pentene and cis-3-hexene

[6] M. F. Hawthorne, *J. Am. Chem. Soc.*, **83**, 2541 (1961).

[7] R. Dulou and Y. Chrétien-Bessière, *Bull soc. chim. France*, 1362 (1959).

[8] A. C. Cope, G. A. Berchtold, P. E. Peterson, and S. H. Sharman, *J. Am. Chem. Soc.*, **82**, 6370 (1960).

[9] T. J. Logan and T. J. Flautt, *J. Am. Chem. Soc.*, **82**, 3446 (1960).

[10] J. Goubeau, R. Epple, D. Ulmschneider, and H. Lehmann, *Angew. Chem.* **67**, 710 (1955).

[11] H. C. Brown and K. J. Murray, *J. Am. Chem. Soc.*, **81**, 4108 (1959).

in yields of 60 to 80 per cent.   The yield of 1-hexene from 1-hexyne was only 7 per cent, confirming the conclusion that this terminal acetylene undergoes dihydroboration preferentially.   The products from the hydroboration of 1-hexyne and 3-hexyne with disiamylborane yielded 1-hexene in a yield of 92 per cent and *cis*-3-hexene in a yield of 90 per cent. These results are summarized in Table 16-2.

Table 16-2

Protonolysis of the Vinylboranes Obtained in the Hydroboration of Some Representative Acetylenes

Acetylene	Hydroborating agent	Olefin formed	Olefin yield, %
1-Hexyne	NaBH$_4$, BF$_3$[a]	1-Hexene	7
1-Hexyne	Sia$_2$BH[b]	1-Hexene	92[c]
2-Pentyne	NaBH$_4$, BF$_3$[a]	*cis*-2-Pentene	60
3-Hexyne	NaBH$_4$, BF$_3$[a]	*cis*-3-Hexene	68
3-Hexyne	NaBH$_4$, BF$_3$[a]	*cis*-3-Hexene	80[c]
3-Hexyne	Sia$_2$BH[b]	*cis*-3-Hexene	90
Diphenylacetylene	Sia$_2$BH[b]	*cis*-Stilbene	69

[a] Internal hydroboration in diglyme.

[b] Hydroboration with disiamylborane in diglyme.

[c] Small excess of hydroborating agent used; otherwise theoretical quantity.

The *cis* olefins prepared in this way exhibit purities in the neighborhood of 98 to 99 per cent.   The conversion of internal acetylenes into *cis* olefins has been previously accomplished by hydrogenation.   The procedure requires careful preparation of the catalyst[12] and purities of no higher than 95 per cent are realized.   Consequently, the hydroboration route appears to offer definite advantages for this transformation.

[12] H. Lindlar, *Helv. Chim. Acta*, **35**, 446 (1952).

A.C. Cope and his co-workers have recently made an interesting application of this procedure by transforming cyclodecyne into *cis*-cyclodecene-1,2-d_2, utilizing hydroboration with hexadeuterodiborane, followed by protonolysis with deuteroacetic acid.[8]

Finally, attention should be called to the related reaction of diisobutyl-aluminum hydride with acetylene, which also provides a convenient conversion into *cis* olefins.[13]

Stereochemistry of the Hydroboration-Protonolysis Reaction

The available evidence indicates that the protonolysis of trialkylboranes proceeds with retention of configuration[14] (Chapter 8). There appears to be no reason to doubt that protonolysis of a vinylborane likewise proceeds with retention. Consequently, the formation of the *cis* olefin must involve a *cis* addition of the boron-hydrogen bond to the acetylene structure, followed by a rapid protonolysis with retention of configuration.

[13] G. Wilke and H. Müller, *Ber.*, **89**, 444 (1956).
[14] H. C. Brown and K. J. Murray, *J. Org. Chem.*, **26**, 631 (1961).

The conclusion that the reaction involves a *cis* addition of the boron-hydrogen bond to the triple bond is confirmed by an infrared examination of the reaction product from disiamylborane and 1-hexyne. The spectrum reveals a strong absorption at 1000 cm^{-1} (10.0 μ),[15] assigned to the out-of-plane deformation vibrations of *trans* vinylic hydrogens.[16]

The addition to the triple bond presumably proceeds through a four-center transition state similar to, that previously proposed for the olefin reaction (Chapter 8).

Oxidation of the Vinyl Organoboranes

The vinyl organoboranes undergo the usual oxidation with alkaline hydrogen peroxide. However, since the products are carbonyl compounds, it is desirable to add the alkali concurrently with the hydrogen peroxide, thereby avoiding strongly alkaline conditions.

Oxidation of the monohydroboration product from 3-hexyne yields 3-hexanone. This is the product formed by the usual hydration procedures, so that hydroboration provides an alternative, but not distinctive,

[15] H. C. Brown and D. Bowman, unpublished observations.

[16] L. J. Bellamy, "The Infra-red Spectra of Complex Molecules," Methuen & Co. Ltd., London, 1958.

procedure for the hydration of internal acetylenes. However, the hydroboration-oxidation of terminal acetylenes results in the formation of aldehydes in excellent yields. Thus, an 88 per cent yield of n-hexaldehyde was realized from 1-hexyne. Since the usual hydration procedures convert terminal acetylenes into methyl ketones, the hydroboration procedures open up a new synthetic path for the utilization of terminal acetylenes.

The reaction apparently involves the usual anti-Markownikoff addition of the boron-hydrogen bond to the terminal acetylene. Oxidation then must proceed through the formation of the enol borate ester (not identified), which is presumably rapidly hydrolyzed to the enol, tautomerizing to the final product, the aldehyde.

Dihydroboration of Acetylenes

In dihydroboration, acetylenes become difunctional molecules. The reaction with the polyfunctional diborane molecule would be expected to produce complex polymers, similar to those encountered in the dihydroboration of dienes.

Treatment of 1- and 3-hexyne with the stoichiometric amount of hydroborating agent at 0° results in an approach to the utilization of 2.0 hydrides per acetylene molecule (1.57 to 1.85), but the reaction fails to go to completion in a reasonable time. Even with a threefold excess of hydride, the hydride up-take is no greater than 1.88 to 1.92 per acetylene molecule. It appears that the reaction must involve the formation of a complex cross-linked polymer with some of the residual double bonds so buried that complete reaction becomes exceedingly slow.

Reaction of an acetylene with diborane could reasonably take three paths: (a) two boron atoms could add to the same carbon atom; (b) two boron atoms could add to adjacent carbon atoms; (c) a double addition of two hydrides from the same boron atom could occur.

It appeared that oxidation with hydrogen peroxide should provide evidence as to the nature of the dihydroboration product. However, oxidation of the product from the dihydroboration of 1-hexyne yielded 1-hexanol as the major product: 10 per cent 1,2-hexanediol, 26 per cent *n*-hexaldehyde, and 64 per cent of 1-hexanol (from a total yield of identified products of 83 per cent).

Logan and Flautt have also observed that the simple alcohol is a major product of the hydroboration-oxidation of di-*t*-butylacetylene.[9]

The only apparent explanation for the high yield of 1-hexanol is that the dihydroboration product, whatever its structure, undergoes rapid hydrolysis to lose one of the two boron-carbon bonds prior to oxidation. On this basis, it appeared that variation of the oxidation procedure might avoid this difficulty. In actual fact, variation in the oxidation procedure had little effect on the product distribution.

It appeared possible that the difficulty might be circumvented by performing the dihydroboration with dicyclohexylborane. After 24 hours at room temperature, nearly two dicyclohexylborane groups per 1-hexyne molecule had reacted.

$$n\text{-}C_4H_9C \equiv\!\equiv\! CH \xrightarrow{\left\langle\right\rangle_2 BH}$$

$$\longrightarrow n\text{-}C_4H_9\underset{\underset{R_2B}{|}}{C}H\underset{\underset{BR_2}{|}}{C}H_2$$

$$\longrightarrow n\text{-}C_4H_9CH_2\underset{\underset{BR_2}{|}}{\overset{\overset{BR_2}{|}}{C}}H$$

Oxidation with alkaline hydrogen peroxide gave an even larger yield of 1-hexanol: 10 per cent n-hexaldehyde and 90 per cent of 1-hexanol, with only traces of the 1,2-diol.

This experiment suggests that this dihydroboration product must be exceedingly unstable to hydrolytic cleavage, in spite of the presence of the large cyclohexyl groups. Assuming that the initial product contains the two boron atoms on the terminal carbon, it is possible to write a reasonable mechanism to account for the hydrolytic instability.

$$RCH_2\underset{\underset{BR_2}{\diagdown}}{\overset{\overset{BR_2}{\diagup}}{C}}H \xrightarrow{\;OH^-\;} RCH_2\overset{\diagup}{\underset{\diagdown}{C}}H \left[\begin{array}{c} HO \\ \overset{\cdot\cdot}{B}R_2 \\ BR_2 \end{array} \right]^{-}$$

$$\downarrow$$

$$RCH_2CH\!=\!\overline{B}R_2 \longleftrightarrow RCH_2\overline{C}HBR_2 + R_2BOH$$

$$\downarrow H_2O$$

$$RCH_2CH_2BR_2$$

Since the initial stage in this hydrolysis involves an attack by base on the electrophilic boron atom, substituents that decrease the acidity of the boron atom should serve to reduce the hydrolytic stability of the 1,1-diboro derivative. It is well known that hydroxy and methoxy substituents greatly decrease the acceptor properties of boron. Accordingly, 1-hexyne was added to a threefold excess of diborane in tetrahydrofuran. The product was hydrolyzed.

$$RCH_2CH \begin{smallmatrix} BH_2 \\ \\ BH_2 \end{smallmatrix} \xrightarrow{\;H_2O\;} RCH_2CH \begin{smallmatrix} B(OH)_2 \\ \\ B(OH)_2 \end{smallmatrix}$$

$$\text{slow} \downarrow \; OH^-$$

$$RCH_2CH_2B(OH)_2 \xleftarrow[\text{slow}]{\;H_2O\;} RCH_2CH \begin{smallmatrix} B(OH)_3^- \\ \\ B(OH)_2 \end{smallmatrix}$$

Oxidation of the product with alkaline hydrogen peroxide at pH 8 yielded 16 per cent 1,2-hexanediol, 68 per cent n-hexaldehyde, and 16 per cent 1-hexanol.

The results clearly indicate that the dihydroboration product contains two boron atoms on the same carbon atom. It is of interest that the analogous reaction of diethyl aluminum hydride with 1-hexyne likewise proceeds to place both of the aluminum atoms on the terminal position.[16]

The formation of 16 per cent of 1,2-hexanediol was observed, corresponding to the formation of a minor quantity of the 1,2-diboro derivative. The hydroboration of terminal olefins proceeds to place 6 to 7 per cent of the boron on the 2 carbon atom (Chapter 7). It is probable that the initial reaction of diborane with a terminal acetylene places an even larger fraction of the boron adding onto the 2 carbon atom of the triple bond. The subsequent addition to place the second boron atom on the more favored 1 position would readily account for the minor amount of the 1,2-diboro derivative apparently present in the product.

From these results, it can be concluded that the dihydroboration of acetylenes occurs preferentially to place two boron atoms on the same carbon atom, with a minor amount of the isomeric derivative. The products are exceedingly sensitive to hydrolytic cleavage, being converted into the corresponding alkylborane. It is evident that the dihydroboration of acetylenes can also be utilized for the synthesis of the corresponding aldehydes or ketones. However, monohydroboration with disiamylborane produces considerably better yields and would appear to be the more satisfactory synthetic procedure.

[16] G. Wilke and H. Müller, *Ann.*, **618**, 267 (1959).

17 | Diborane as a Reducing Agent

The discussion thus far has dealt largely with applications of the hydroboration reaction to simple hydrocarbons — olefins, dienes, and acetylenes. Important as these simple compounds are, the hydroboration reaction is by no means limited to this class of substances. Diborane, although a strong reducing agent,[1-3] appears to react with representative carbon-carbon double and triple bonds far more rapidly than it does with most of the functional groups.[4] Consequently, the selective hydroboration of unsaturated molecules containing such functional groups would appear quite practical, and has, indeed, been demonstrated in a modest number of cases (Chapter 19).

The organoboranes are relatively reactive intermediates, capable of numerous transformations of interest in organic synthesis. The presence of most functional groups cannot be tolerated in the more usual organometallics commonly used in organic synthesis, such as the organolithium and organomagnesium compounds. For this reason the hydroboration of olefinic and acetylenic molecules containing functional groups is of special interest and promise.

In order to apply the hydroboration reaction to such functional derivatives, the chemist should be familiar with the reactions of diborane as a reducing agent, as well as with the action of sodium borohydride and related derivatives, since the hydroboration procedures involve placing the compound undergoing hydroboration in solution with one or more of these reagents. For that reason it has appeared desirable in this

[1] H. C. Brown, H. I. Schlesinger, and A. B. Burg, *J. Am. Chem. Soc.*, **61**, 673 (1939).
[2] H. C. Brown and B. C. Subba Rao, *J. Org. Chem.*, **22**, 1135 (1957).
[3] H. C. Brown and B. C. Subba Rao, *J. Am. Chem. Soc.*, **82**, 681 (1960).
[4] H. C. Brown and W. Korytnyk, *J. Am. Chem. Soc.*, **82**, 3866 (1960).

238

chapter to discuss the reducing powers of diborane, as well as to review briefly the reducing action of sodium borohydride and related derivatives.[5]

It was pointed out earlier that the use of disiamylborane makes possible many selective hydroborations (Chapter 13). Disiamylborane is also a reducing agent, but its action is much milder than that of diborane.[6] In some cases the use of this substituted borane offers major advantages for the hydroboration of functional derivatives. Consequently, its behavior will be reviewed in Chapter 18 before we turn our attention to a consideration of the hydroboration of functional derivatives (Chapter 19).

Lithium Aluminum Hydride as a Reducing Agent

Lithium aluminum hydride is readily soluble in the usual hydroboration solvents — ethyl ether, tetrahydrofuran, and diglyme. In all these solvents it is a powerful reducing agent, generally reducing such groups rapidly to the lowest reduced state.[7] This is indicated by the following list of representative groups, arranged approximately in the order of their relative ease of reduction by the complex hydrides.

$$
\begin{aligned}
&\text{Aldehyde} \longrightarrow \text{alcohol} \\
&\text{Ketone} \longrightarrow \text{alcohol} \\
&\text{Acid chloride} \longrightarrow \text{alcohol} \\
&\text{Lactone} \longrightarrow \text{glycol} \\
&\text{Oxide} \longrightarrow \text{alcohol} \\
&\text{Ester} \longrightarrow \text{alcohol} \\
&\text{Carboxylic acid} \longrightarrow \text{alcohol} \\
&\text{Carboxylic acid salt} \longrightarrow \text{alcohol} \\
&\text{Nitrile} \longrightarrow \text{amine} \\
&\text{Nitro} \longrightarrow \text{azo, etc.} \\
&\text{Olefin} \longrightarrow \text{no reaction}
\end{aligned}
$$

Consequently, it would appear that the use of this reagent would have little scope for the hydroboration of olefinic and acetylenic molecules containing functional groups.[8] Ether and tertiary amine groups should be accommodated. Hydroxy and amine groups evolve hydrogen with

[5] H. C. Brown, *J. Chem. Educ.*, **38,** 173 (1961).

[6] H. C. Brown and D. B. Bigley, *J. Am. Chem. Soc.*, **83,** 486 (1961).

[7] W. G. Brown, Reductions by Lithium Aluminum Hydride, "Organic Reactions," Vol. 6, Chap. 10, p. 469, John Wiley & Sons, Inc., New York, 1951.

[8] F. Sondheimer and S. Wolfe, *Canadian J. Chem.*, **37,** 1870 (1959).

the reagent but should cause no difficulty otherwise. Relatively reactive halides are attacked by lithium aluminum hydride, but the more stable derivatives should not be affected by this compound.

The reactivity of lithium aluminum hydride can be influenced markedly by alkoxy substituents. For example, treatment of lithium aluminum hydride with 3 moles of *t*-butyl alcohol forms lithium tri-*t*-butoxyaluminohydride.[9]

$$LiAlH_4 + 3(CH_3)_3COH \longrightarrow LiAlH[OC(CH_3)_3]_3 + 3H_2$$

This reagent is much milder than the parent compound. It reduces aldehydes, ketones, and acid chlorides but fails to reduce other functional groups under the usual mild conditions (0 to 25°) at any significant rate.

Aldehyde $\longrightarrow$ alcohol
Ketone $\longrightarrow$ alcohol
Acid chloride $\longrightarrow$ aldehyde
Acid chloride $\longrightarrow$ alcohol
Ester $\longrightarrow$ no reaction
Nitrile $\longrightarrow$ no reaction
Olefin $\longrightarrow$ no reaction

The reagent is particularly valuable for the partial reduction of acid chlorides to aldehydes.[9,10] For this purpose the acid chloride in diglyme or tetrahydrofuran at low temperatures (preferably −78°) is treated with one equivalent of the reagent in the same solvent. The temperature is permitted to rise to room temperature, the reaction mixture poured onto crushed ice, and the aldehyde isolated. In this way *p*-nitrobenzoyl or terphthalyl chlorides can be converted to the corresponding aldehydes in yields of 80 to 90 per cent.

Lithium trimethoxyaluminohydride can be synthesized by an analogous reaction.

$$LiAlH_4 + 3CH_3OH \longrightarrow LiAlH(OCH_3)_3 + 3H_2$$

It resembles lithium aluminum hydride in its reducing power.[11]

[9] H. C. Brown and R. F. McFarlin, *J. Am. Chem. Soc.*, **80**, 5372 (1958).
[10] H. C. Brown and B. C. Subba Rao, *J. Am. Chem. Soc.*, **80**, 5377 (1958).
[11] H. C. Brown and C. J. Shoaf, manuscript in preparation.

Aldehyde ⟶ alcohol
Ketone ⟶ alcohol
Acid chloride ⟶ alcohol
Lactone ⟶ glycol
Oxide ⟶ alcohol
Ester ⟶ alcohol
Carboxylic acid ⟶ alcohol
Nitrile ⟶ aldehyde
Nitrile ⟶ amine

The presence of the single active hydride on the anion, instead of the four on lithium aluminum hydride, makes possible a number of interesting selective reductions. Thus the related derivatives, sodium[12,13] and lithium[11,14] triethoxyaluminohydrides, have been utilized for the reduction of nitriles to aldehydes.

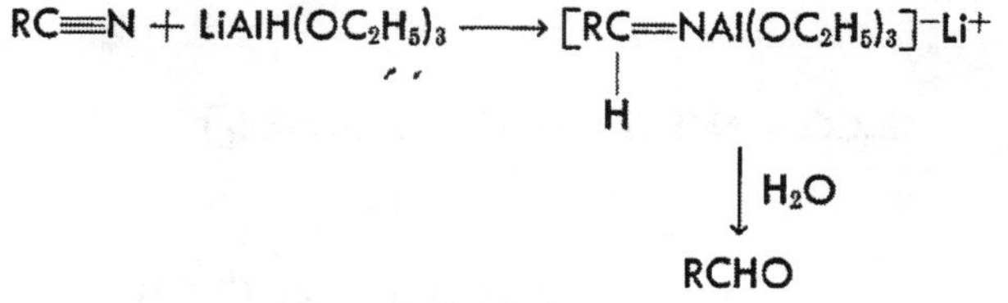

$$RC{\equiv}N + LiAlH(OC_2H_5)_3 \longrightarrow [RC{=}NAl(OC_2H_5)_3]^- Li^+$$
$$\underset{H}{|}$$

$$\downarrow H_2O$$

$$RCHO$$

Similarly, lithium triethoxyaluminohydride converts acyldimethylamides into aldehydes in high yields.[15]

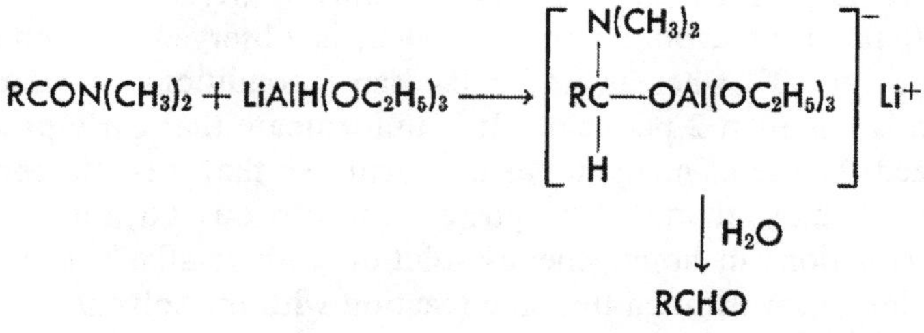

$$RCON(CH_3)_2 + LiAlH(OC_2H_5)_3 \longrightarrow \left[\begin{array}{c} N(CH_3)_2 \\ | \\ RC{-}OAl(OC_2H_5)_3 \\ | \\ H \end{array} \right]^- Li^+$$

$$\downarrow H_2O$$

$$RCHO$$

The reducing power of some of these modified reagents is relatively mild, approaching sodium borohydride in activity. Presumably, they could be utilized as hydride sources for the hydroboration of functional

[12] O. Schmitz-DuMont and V. Habernickel, *Ber.*, **90**, 1054 (1957).
[13] G. Hesse and R. Schrödel, *Ann.*, **607**, 24 (1957).
[14] H. C. Brown, C. J. Shoaf, and C. P. Garg, *Tetrahedron Letters*, No. 3, 9 (1959).
[15] H. C. Brown and A. Tsukamoto, *J. Am. Chem. Soc.*, **81**, 502 (1959).

derivatives and would allow a considerably larger scope than lithium aluminum hydride itself. However, for such hydroborations they do not appear to have any advantage over sodium borohydride itself, and no case of their utilization for this purpose has appeared in the literature.

Sodium Borohydride as a Reducing Agent

Sodium borohydride has the great advantage over lithium aluminum hydride as a reducing agent in that the borohydride may be utilized in a much wider range of solvents. Thus it is highly soluble in water, reacting only very slowly with the solvent to evolve hydrogen. Even this slow reaction is markedly decreased by the addition of alkali,[16] and stable solutions of sodium borohydride in strong caustic are commercially available. Such aqueous solvents readily reduce aldehydes and ketones, even in cases in which the solubilities of the compounds in the aqueous system are quite limited.

$$4R_2CO + NaBH_4 \longrightarrow Na^+[B(OCHR_2)_4]^-$$
$$\downarrow 4H_2O$$
$$NaB(OH)_4 + 4R_2CHOH$$

The reducing agent is readily soluble in methyl and ethyl alcohols.[16] It reacts rapidly with methyl alcohol but only slowly with ethyl alcohol. Thus 100 per cent evolution of hydrogen is observed in methanol in 24 minutes at 60°, whereas under the same conditions in ethanol the evolution is less than 2 per cent. It is unfortunate that early procedures emphasized the use of methanol as a solvent[17] so that it is still commonly applied, although ethanol does possess the obvious advantage of permitting reductions in homogeneous solution with relatively little loss of the reducing agent through this side reaction with the solvent.

Sodium borohydride possesses only a modest solubility (0.1 M at 25°) in isopropyl alcohol. However, this solvent possesses the great advantage of being quite stable toward the reagent; standard solutions of sodium borohydride in this solvent have exhibited no measurable change in active hydride over periods of several weeks. Consequently, this solvent not only

[16] H. I. Schlesinger, H. C. Brown, H. R. Hoekstra, and L. R. Rapp, *J. Am. Chem. Soc.*, **75**, 199 (1953).

[17] S. W. Chaiken and W. G. Brown, *J. Am. Chem. Soc.*, **71**, 122 (1949).

is convenient for reductions on a preparative scale, but it made possible kinetic studies of the reduction of aldehydes and ketones.[18–20]

The very great differences in reactivity between benzaldehyde and acetophenone observed in these kinetic studies is noteworthy. Thus the second-order rate constant for acetophenone at 0° is 2.05×10^{-4} liter mole^{-1} sec^{-1}, whereas the corresponding constant for benzaldehyde is 820×10^{-4} liter mole^{-1} sec^{-1}. With a factor of 400 in rate constants, it is evident that aldehyde groups should be reduced selectively in the presence of related ketone groups.

Although sodium borohydride in hydroxylic solvents does reduce aldehyde and ketone groups rapidly at 25°, it is essentially inert to the other groups except under more vigorous conditions or unusual structures.

Aldehyde ⟶ alcohol
Ketone ⟶ alcohol
Acid chloride ⟶ reaction with solvent
Lactone ⟶ no reaction
Oxide ⟶ no reaction
Ester ⟶ no reaction
Carboxylic acid ⟶ reacts, no reduction
Carboxylic acid salt ⟶ no reaction
Nitrile ⟶ no reaction
Nitro ⟶ no reaction
Olefin ⟶ no reaction

The hydroboration reaction requires acidification of the reaction mixture containing the unsaturated component and sodium borohydride. Under these conditions sodium borohydride reacts rapidly with the liberation of hydrogen and the formation of the borate ester.

$$3ROH + NaBH_4 + HCl \longrightarrow (RO)_3B + 4H_2 + NaCl$$

This reaction provides a simple synthesis of many borate esters not easily obtained otherwise,[21] but renders these solvents unsuitable for hydroborations.

[18] H. C. Brown, O. H. Wheeler, and K. Ichikawa, *Tetrahedron*, 1, 214 (1957).
[19] H. C. Brown and K. Ichikawa, *Tetrahedron*, 1, 221 (1957).
[20] H. C. Brown and K. Ichikawa, *J. Am. Chem. Soc.*, 84, 373 (1962).
[21] H. C. Brown, E. J. Mead, and C. J. Shoaf, *J. Am. Chem. Soc.*, 78, 3613 (1956).

Sodium borohydride is insoluble in ethyl ether, only slightly soluble in tetrahydrofuran, but readily soluble in diglyme and triglyme.[22] Consequently, diglyme and triglyme have been the solvents of choice for hydroborations based on sodium borohydride.

In diglyme, sodium borohydride reduces aldehydes, but the reaction with ketones is very slow.   Acid chlorides are rapidly reduced.   Carboxylic acids react to liberate hydrogen and form an acylborohydride, but no reduction occurs.   Other groups examined are essentially inert under the experimental conditions examined (0 to 25°).[22,23]

Aldehyde ⟶ alcohol
Ketone ⟶ very slow reaction
Acid chloride ⟶ alcohol
Lactone ⟶ no reaction
Oxide ⟶ no reaction
Ester ⟶ no reaction
Carboxylic acid ⟶ reacts, no reduction
Carboxylic acid salt ⟶ no reaction
Nitrile ⟶ no reaction
Nitro ⟶ no reaction
Olefin ⟶ no reaction

It was early noted that there is a significant difference in the reducing power of lithium and sodium borohydrides.   Whereas sodium borohydride reduces typical esters (such as ethyl acetate and ethyl benzoate) only very slowly, lithium borohydride reduces such esters quite easily.[24]   In aqueous solution there is no measurable difference in the rate constants for the reactions of sodium and lithium borohydrides with acetone.   However, in isopropyl alcohol solution the rate constant for lithium borohydride is several times greater than that for the sodium salt.[25]   This suggests that the enhancing effect of the lithium ion will be greatest in solvents of low dielectric constant.   In such solvents the reaction presumably proceeds through the ion pair, ($Li^+BH_4^-$), rather than the dissociated ions.

Kollonitsch and his co-workers achieved the successful reduction of

[22] H. C. Brown, E. J. Mead, and B. C. Subba Rao, J. Am. Chem. Soc., 77, 6209 (1955).

[23] H. C. Brown and B. C. Subba Rao, unpublished observations.

[24] R. F. Nystrom, S. W. Chaiken, and W. G. Brown, J. Am. Chem. Soc., 71, 3245 (1949).

[25] H. C. Brown and K. Ichikawa, J. Am. Chem. Soc., 83, 4372 (1961).

simple esters by sodium borohydride in tetrahydrofuran in the presence of lithium and magnesium iodides.[26]   Later they synthesized calcium, strontium, and barium borohydrides, and demonstrated that these borohydrides exhibit an enhanced reducing power over sodium borohydride.[27]

In our own approach to this problem, we utilized homogeneous solutions of sodium borohydride in diglyme. Thus the addition of an equivalent quantity of lithium chloride or lithium bromide to a 1.0 M solution of sodium borohydride in diglyme results in the formation of a precipitate of sodium halide and the formation *in situ* of lithium borohydride. The reagent can be utilized directly, without removing the precipitated salt. With this reagent, esters were reduced, slowly at room temperature, rapidly and quantitatively at higher temperatures. The enhanced reducing power of the reagent is indicated by the results with the standard compounds.

Aldehyde $\longrightarrow$ alcohol

Ketone $\longrightarrow$ alcohol

Acid chloride $\longrightarrow$ alcohol

Lactone $\longrightarrow$ glycol

Oxide $\longrightarrow$ alcohol

Ester $\longrightarrow$ alcohol

Carboxylic acid $\longrightarrow$ reacts, no reduction

Carboxylic acid salt $\longrightarrow$ no reaction

Nitrile $\longrightarrow$ no reaction

Nitro $\longrightarrow$ no reaction

Olefin $\longrightarrow$ no reaction

Anhydrous magnesium chloride and bromide possess only a small solubility in diglyme (0.013 M for magnesium chloride and 0.078 M for the bromide, both at 100°). Although their addition to the diglyme solutions of sodium borohydride has a considerable effect, the insolubility of the salts represents an inconvenient feature.[22] On the other hand, anhydrous aluminum chloride is soluble in diglyme and a solution of this material with sodium borohydride in the molar ratio of 1:3 provides a clear solution. Nevertheless, such solutions exhibit markedly enhanced reducing powers.[28]

[26] J. Kollonitsch, O. Fuchs, and V. Gabor, *Nature*, **173**, 125 (1954).

[27] J. Kollonitsch, O. Fuchs, and V. Gabor, *Nature*, **175**, 346 (1955).

[28] H. C. Brown and B. C. Subba Rao, *J. Am. Chem. Soc.*, **78**, 2582 (1956).

At room temperatures, such solutions rapidly reduce most of the groups in the standard list.

Aldehyde ⟶ alcohol
Ketone ⟶ alcohol
Acid chloride ⟶ alcohol
Lactone ⟶ glycol
Oxide ⟶ alcohol
Ester ⟶ alcohol
Carboxylic acid ⟶ alcohol
Carboxylic acid salt ⟶ no reaction
Nitrile ⟶ amine
Nitro ⟶ no reaction
Olefin ⟶ organoborane

In addition to the standard list, it might be mentioned that tertiary amides, azo compounds, and disulfides also undergo rapid reduction. No reduction was observed for primary amides, sulfones, and polycyclic aromatics, in addition to the nitro compounds and sodium carboxylates indicated in the list.

As was pointed out earlier (Chapter 2), it was the observation that this reagent reacted with olefins which led to the entire hydroboration program.[29]

It is interesting that alkoxy substituents increase the reducing power of sodium borohydride,[30] whereas they decrease the reducing power of lithium aluminum hydride.[9] This difference in behavior has been attributed to the marked difference in the resonance interactions of the alkoxy substituent with boron and with aluminum.[9]

Reductions by these complex hydrides presumably proceed through a transfer of hydride ion from the anion to the groups undergoing reduction.

$$
\begin{array}{cc}
\overset{\displaystyle H}{\underset{\displaystyle H}{H-B}}---\overset{-}{H}---\overset{\displaystyle}{C}{=}O
&
\overset{\displaystyle OR}{\underset{\displaystyle OR}{RO-B}}---\overset{-}{H}---\overset{\displaystyle}{C}{=}O
\end{array}
$$

On this basis, the transfer of hydride ion from the weaker Lewis acid, alkyl borate, should proceed more rapidly than the transfer from the stronger Lewis acid, borane.

[29] H. C. Brown and B. C. Subba Rao, *J. Am. Chem. Soc.*, **78**, 5694 (1956).
[30] H. C. Brown, E. J. Mead, and C. J. Shoaf, *J. Am. Chem. Soc.*, **78**, 3616 (1956).

Unfortunately, little is known about the relative strengths of aluminum hydride and aluminum alkoxides as Lewis acids. The polymeric nature · of the products renders difficult the experimental comparison of their acid strengths. Fortunately, it is possible to rationalize the observed effect on theoretical grounds.

In the case of both the alkyl borates and the aluminum alkoxides, the electron-withdrawing inductive effect of the alkoxy groups would be expected to increase the acid strengths of the two related derivatives, $(RO)_3B$ and $(RO)_3Al$. However, large resonance contributions in the alkyl borate satisfy the electron deficiency of the boron atom and reverse the anticipated electronic effect of the alkoxy groups.

Consequently, instead of being strong Lewis acids, alkyl borates are quite weak.

Such resonance interactions should be far less important in a second-row element, such as aluminum, than in a first-row element, such as boron. It would not be unexpected if the electronic effect of the alkoxy groups in aluminum derivatives were primarily a reflection of the electron-withdrawing inductive effect of these groups with only minor modifications by the less important resonance interactions. In this way the markedly different effects of alkoxy substituents on the reducing powers of the borohydride and aluminohydride derivatives are the result of the different capacity of boron and aluminum to participate in resonance interactions involving double-bonded structures with the oxygen atom of the substituent.

Diborane as a Reducing Agent

Reductions by alkali metal borohydrides and aluminohydrides appear to involve a transfer of hydride ion from the anion to an electron-deficient center of the functional group.[31] Consequently, the introduction of electron-withdrawing substituents would be expected to result in an increase in the rate of reduction. This is observed. For example, both

[31] L. W. Trevoy and W. G. Brown, *J. Am. Chem. Soc.*, **71**, 1675 (1949).

chloral and acetyl chloride are much more rapidly reduced in diglyme than simple aldehydes and ketones.

$$
\begin{array}{ccc}
& CH_3 & H \\
& | & | \\
CH_3-C&-----&C=O \\
& | \\
& CH_3
\end{array}
\qquad
\begin{array}{ccc}
& Cl & H \\
& | & | \\
Cl-C&-&C=O \\
& | \\
& Cl
\end{array}
$$

↑
preferential
attack by
NaBH₄

On the other hand, diborane is a strong Lewis acid. It would be expected to involve a preferred electrophilic attack on the centers of highest electron densities. Indeed, diborane reduces trimethylacetaldehyde readily, but fails to react with chloral under the usual mild conditions (0 to 25°).[1]

$$
\begin{array}{ccc}
& CH_3 & H \\
& | & | \\
CH_3-C&-----&C=O \\
& | \\
& CH_3
\end{array}
\qquad
\begin{array}{ccc}
& Cl & H \\
& | & | \\
Cl-C&-&C=O \\
& | \\
& Cl
\end{array}
$$

↑
preferential
attack by
B₂H₆

It was evident that a reagent with such different reducing characteristics might be exceedingly useful in selective reductions, and an extensive study of its behavior was therefore made.[2] Because of the high solubility of diborane in tetrahydrofuran, such solutions are conveniently utilized for reductions by diborane. Alternatively, it is often possible to generate diborane *in situ* from the reaction of sodium borohydride and boron trifluoride etherate.[3]

Both aliphatic and aromatic ketones are rapidly reduced at room temperature. However, chloral is inert to the reagent. Such reductions generally involve the rapid reaction of 2 moles of carbonyl compound per mole of borane to form the dialkoxyborane. The third hydride reacts with difficulty. Hydrolysis yields the alcohol.

$$2RCHO + BH_3 \longrightarrow (RCH_2O)_2BH$$

$$(RCH_2O)_2BH + 3H_2O \longrightarrow 2RCH_2OH + B(OH)_3 + H_2$$

Both lactones and oxides reduce readily, but the reactions are considerably slower than those of aldehydes and ketones. Esters are reduced, but slower still. On the other hand, acid chlorides fail to react. Presumably, the electron-withdrawing influence of the chlorine substituent has an effect on the carbonyl group similar to that of the three chlorine substituents in chloral. Nitro compounds do not react.

Perhaps the most interesting observation is that both carboxylic acids and nitriles are rapidly reduced by the reagent. Indeed, carboxylic acids are reduced considerably faster than ketones and perhaps as fast or faster than aldehydes. On the other hand, sodium carboxylates do not react.

On this basis the results with the standard substrates are as follows.

Aldehyde $\longrightarrow$ alcohol
Ketone $\longrightarrow$ alcohol
Acid chloride $\longrightarrow$ no reaction
Lactone $\longrightarrow$ glycol
Oxide $\longrightarrow$ alcohol
Ester $\longrightarrow$ alcohol (slow)
Carboxylic acid $\longrightarrow$ alcohol (fast)
Nitrile $\longrightarrow$ amine
Nitro $\longrightarrow$ no reaction
Olefin $\longrightarrow$ organoborane (fast)

With regard to other groups, it may be noted that primary amides react to evolve 2 moles of hydrogen but do not reduce, sulfones are not reduced, but sulfoxides undergo a slow reaction. Neither aryl nor alkyl halides react. Polycyclic aromatics, including anthracene, appear stable to the reagent under the experimental conditions. Azo compounds are reduced to amines, and it has been recently noted that oximes are reduced to the corresponding hydroxylamines.[32]

[32] H. Feuer and B. F. Vincent, Purdue University, unpublished observation.

A study of the relative reactivity of a number of representative groups toward diborane indicates the following order of reactivity[4]: carboxylic acids > olefins > ketones > nitriles > epoxides > esters > acid chlorides.

On the other hand, toward alkali metal borohydride the order is: acid chlorides > ketones > epoxides > esters > nitriles > carboxylic acids.

With such markedly different reactivities, it is apparent that the judicious use of either diborane or alkali metal borohydride permits the reduction of one group in the presence of a second, or the reverse operation.

It is also apparent from this reactivity series that it should be possible to achieve hydroboration of double or triple carbon-carbon bonds with diborane in the presence of most of the functional groups examined. Only carboxylic acid and aldehyde groups possess reactivities that would be competitive with the hydroboration reaction. Fortunately, the carboxylic acid group can be protected merely by forming the ester, the acid chloride, or even the sodium salt; aldehyde groups can be protected as the acetals.

A word of caution is in order. The reactivity of any functional group can be modified greatly by the organic structure to which it is attached. For example, hydroboration of the carbon-carbon double bond appears to be a reaction of exceptional generality. Yet it is possible so to bury a double bond within the steroid structure that the reaction proceeds exceedingly slowly, if at all (Chapter 6). Similarly, aldehydes are generally highly reactive toward diborane, yet the three chlorine substituents in chloral make this otherwise reactive group quite inert toward diborane. It is therefore essential to recognize that the relative reactivities established by the present study must be considered approximate values for simple representative groups and may be greatly altered or even inverted by major modifications in the molecular structure.

Theoretical Considerations

Certain of the unusual characteristics of the reducing action of diborane warrant further analysis.

It has been pointed out that the carbonyl groups of simple aldehydes and ketones react rapidly with diborane, whereas the carbonyl groups of chloral and acid chlorides are relatively inert to the reagent. It was considered significant that those aldehydes and ketones which react readily with diborane also form stable addition compounds with boron trifluoride, whereas chloral and acetyl chloride add this Lewis acid only with difficulty at low temperatures.

$$\text{(CH}_3)_3\text{C}\overset{\text{H}}{\text{C}}=\text{O} + \text{BF}_3 \xrightleftharpoons{\hspace{1cm}} \text{(CH}_3)_3\text{C}\overset{\text{H}}{\text{C}}=\overset{+}{\text{O}}:\overset{-}{\text{BF}}_3$$

$$\text{Cl}_3\text{C}\overset{\text{H}}{\text{C}}=\text{O} + \text{BF}_3 \xrightleftharpoons{\hspace{1cm}} \text{Cl}_3\text{C}\overset{\text{H}}{\text{C}}=\overset{+}{\text{O}}:\overset{-}{\text{BF}}_3$$

$$\text{H}_3\text{C}\overset{\text{Cl}}{\text{C}}=\text{O} + \text{BF}_3 \xrightleftharpoons{\hspace{1cm}} \text{H}_3\text{C}\overset{\text{Cl}}{\text{C}}=\overset{+}{\text{O}}:\overset{-}{\text{BF}}_3$$

It was suggested that the initial stage of the reaction of diborane with the carbonyl group involves a similar acid-base interaction, with borane adding to the oxygen atom, followed by transfer of a hydride unit from boron to carbon.[1]

$$-\overset{|}{\text{C}}=\text{O} + \tfrac{1}{2}(\text{BH}_3)_2 \xrightleftharpoons{\hspace{1cm}} -\overset{|}{\text{C}}=\overset{+}{\text{O}}:\overset{-}{\text{BH}}_3$$

$$\downarrow$$

$$-\overset{|}{\underset{\text{H}}{\text{C}}}-\text{OBH}_2$$

On this basis, the inertness of acetyl chloride and of chloral toward diborane was attributed to the decreased basic properties of the oxygen atom of the carbonyl group resulting from the powerful inductive effects of the halogen substituents.

$$\text{Cl}\!\leftarrow\!\overset{\overset{\text{Cl}}{\uparrow}}{\underset{\downarrow}{\underset{\text{Cl}}{\text{C}}}}-\overset{\text{H}}{\text{C}}\overset{\delta+}{=}\text{O} + \tfrac{1}{2}(\text{BH}_3)_2 \xrightleftharpoons{\hspace{1cm}} \text{Cl}_3\text{C}\overset{\text{H}}{\text{C}}=\overset{+}{\text{O}}:\overset{-}{\text{BH}}_3$$

$$\text{CH}_3\overset{\overset{\text{Cl}}{\uparrow}}{\text{C}}\overset{\delta+}{=}\text{O} + \tfrac{1}{2}(\text{BH}_3)_2 \xrightleftharpoons{\hspace{1cm}} \text{CH}_3\overset{\text{Cl}}{\text{C}}=\overset{+}{\text{O}}:\overset{-}{\text{BH}}_3$$

It is consistent with this interpretation that both acetyl chloride and chloral are highly reactive toward sodium borohydride, much more reactive than simple aldehydes and ketones. The electron deficiency at the carbonyl group, which serves to inhibit attack by the Lewis acid diborane, would be expected to facilitate attack by the nucleophilic borohydride ion.

In other words, diborane is a Lewis acid which functions best as a reducing agent in attacking groups at positions of high electron density, whereas borohydride is a Lewis base which prefers to attack functional groups at positions of low electron density.

This interpretation can be extended to account for the behavior of nitriles. The nitrile group is relatively insensitive to attack by nucleophilic reagents. Consequently, its relative inertness toward attack by borohydride is not unexpected. However, the nitrogen atom of the nitrile group is relatively basic. It adds boron trifluoride to form addition compounds of moderate stability. Presumably, the rapid reduction of nitriles by diborane involves an initial attack of the reagent at this relatively basic position.

Two factors presumably account for the relatively slow rate of reduction of carboxylic esters. First, addition of the borane particle to the oxygen atom of the carbonyl group must compete with addition to the oxygen atom of the alkoxy group. Second, transfer of hydride from boron to carbon will presumably be hindered by the stabilization provided the carbonyl group by resonance with the oxygen atom of the alkoxy group.

$$\begin{array}{ccc} \overset{-}{B}H_3 & & \overset{-}{B}H_3 \\ \overset{+}{O} & & O \\ \parallel & & \mid \\ -C-OR & \longleftrightarrow & -C=\overset{+}{O}R \end{array}$$

The very fast rate of reaction of diborane with carboxylic acids remains to be considered. The first stage in this reaction appears to be the formation of a triacylborane.

$$3RCO_2H + \tfrac{1}{2}(BH_3)_2 \longrightarrow (RCO_2)_3B + 3H_2$$

This initial reaction product is readily reduced by further treatment with diborane. Indeed, the carboxylic acid group in this intermediate is so active that reduction by sodium borohydride is observed also.

The following interpretation of this interesting phenomenon is proposed. The electron deficiency of a boron atom in the triacylborane should exert a powerful demand on the electron pairs of the acyl oxygen. Such resonance as occurs will involve interaction of this oxygen atom with the boron atom, rather than the usual resonance with the carbonyl group.

$$\underset{\overset{\|}{O}}{-}\overset{O}{\underset{}{C}}-O-B{\Big\langle} \longleftrightarrow \underset{\overset{\|}{O}}{-}\overset{O}{\underset{}{C}}-\overset{+}{O}=\overset{-}{B}{\Big\langle}$$

According to this interpretation, the carbonyl groups in the triacylboranes should resemble those in aldehydes and ketones far more than the less active carbonyl groups of esters.

From this summary (see Table 17-1), it is evident that sodium borohydride will permit the reduction of an acid chloride group in the presence of

Table 17-1

Summary of Behavior of Various Functional Groups
toward the Hydride Reagents

	$NaBH_4$ in ethanol	$NaBH_4$ + LiCl in diglyme	$NaBH_4$ + $AlCl_3$ in diglyme	$NaBH_4$ + BF_3 in diglyme	B_2H_6 in tetrahydrofuran	Bis-3-methyl-2-butylborane in THF	$LiAlH(O-t-Bu)_3$ in THF	$LiAlH(OMe)_3$ in THF	$LiAlH_4$ in ether
Aldehyde	+	+	+	+	+	+	+	+	+
Ketone	+	+	+	+	+	+	+	+	+
Acid chloride	+	+	+	+a	−	−	+e	+	+
Lactone	+	+	+	+	+	+c		+	+
Oxide	−	+	+	+	+	−		+	+
Ester	−	+	+	±b	±b	−	−	+	+
Carboxylic acid	−	−	+	+	+	−d		+	+
Carboxylic salt	−	−	−	−	−				+
Nitrile	−	−	+	+	+	+	−	+	+
Nitro	−	−	−	−	−	+			+
Olefin	−	−	+	+	+	+	−	−	−

a Fast reaction with the sodium borohydride prior to addition of the boron trifluoride.

b Moderate reaction with longer reaction times.

c Reaction in a 1:1 ratio.

d Reaction to form derivative, RCO_2BR_2, but no reduction.

e Yields aldehyde with reverse addition.

an ester grouping, whereas diborane will reduce the ester in the presence of the acid chloride grouping. Similarly, lithium borohydride will reduce the ester group in the presence of the nitrile group, whereas diborane will do the opposite. With these reagents it is simple to reduce a carboxylic acid group without attack on a nitro substituent. The possibilities for numerous other selective reductions is indicated by the summary.

The successful hydroboration of functional derivatives involves the selection of reagents that will react with the olefinic or acetylenic bond without reducing the functional group. It is evident that hydroboration of the carbon-carbon multiple bonds by diborane should proceed without difficulty in the presence of nitro, ester, and acid chloride groups. Similarly, sodium borohydride–boron trifluoride should serve to hydroborate unsaturated nitro and ester derivatives. However, the fast reaction of sodium borohydride with acid chlorides would appear to rule out this reagent for the hydroboration of unsaturated acid chlorides.

A more detailed discussion of the selective hydroboration of functional derivatives will be presented in Chapter 19.

18 | Disiamylborane as a Reducing Agent

For selective reactions, which involve careful control of reaction conditions to achieve the preferred alteration of one group in the presence of a less reactive related group, there are obvious disadvantages in the use of polyfunctional reagents. With such a reagent, each reaction stage produces a new reagent with different properties. It is difficult to define optimum reaction conditions for a truly selective reduction under these circumstances.

This difficulty was overcome, in the case of lithium aluminum hydride, by introducing three alkoxy substituents.

$$LiAlH_4 + 3ROH \longrightarrow Li^+[AlH(OR)_3]^-$$

The resulting reagents then proved very valuable for the selective reduction of acid chlorides,[1] acyldimethylamides,[2] and nitriles[3] to the corresponding aldehydes (Chapter 17).

Similarly, in achieving the selective hydroboration of olefins,[4] dienes,[5] and acetylenes[6] there proved to be major advantages in utilizing the dialkylborane, disiamylborane, instead of the trifunctional borane (Chapters 13, 15, and 16). It therefore appeared desirable to explore the utility of a dialkylborane for selective reductions and disiamylborane was selected for the first of such studies.[7]

[1] H. C. Brown and R. F. McFarlin, *J. Am. Chem. Soc.*, **80**, 5372 (1958).
[2] H. C. Brown and A. Tsukamoto, *J. Am. Chem. Soc.*, **81**, 502 (1959).
[3] H. C. Brown, C. J. Shoaf, and C. P. Garg, *Tetrahedron Letters*, No. 3, 9 (1959).
[4] H. C. Brown and G. Zweifel, *J. Am. Chem. Soc.*, **83**, 1241 (1961).
[5] G. Zweifel, K. Nagase, and H. C. Brown, *J. Am. Chem. Soc.*, **84**, 190 (1962).
[6] H. C. Brown and G. Zweifel, *J. Am. Chem. Soc.*, **83**, 3834 (1961).
[7] H. C. Brown and D. B. Bigley, *J. Am. Chem. Soc.*, **83**, 486 (1961).

Scope

In the initial exploration of the behavior of disiamylborane as a reducing agent, a standard solution of the reagent in tetrahydrofuran, approximately 1.0 M, was mixed with the representative compound, and the reaction mixture was maintained at 0°. An excess of the reducing agent was used — 4 moles of disiamylborane per mole of substrate. The amount of hydrogen evolved in the course of the reaction was noted (Table 18-1). At 14 and 38 hours identical reaction mixtures were analyzed for residual

Table 18-1

Reaction of Disiamylborane with Representative Organic Derivatives in Tetrahydrofuran at 0°[a]

Compound	Hydrogen evolved[b]	Hydride utilized for reduction[b]	Product
Alcohol	1.0	0	Alcohol
Phenol	1.0	0	Phenol
Aldehyde	0	1.0	Alcohol
Ketone	0	1.0	Alcohol
Epoxide	0	0.1	Slight reduction
Carboxylic acid	1.0	0	Carboxylic acid
Acid chloride	0	0	No reaction
Acid anhydride	0	0	No reaction
Ester	0	0	No reaction
Lactone	0	1.0	Hydroxyaldehyde
Enol acetate	0	2.0	
Acid amide	2.0	0	Acid amide
N,N-dimethylamide	0	1.0	Aldehyde
Oxime	1.0	0	Oxime
Nitrile	0	[c]	Slow reduction
Azo	0	0	No reaction
Nitro	0	[c]	Slow reduction
Thiol	0	0	No reaction
Disulfide	0	0	No reaction
Sulfone	0	0	No reaction
Sulfonic acid hydrate	3.0	0	Sulfonic acid
Olefin	0	1.0	Organoborane
Acetylene	0	1.0	Vinylborane

[a] Disiamylborane was approximately 1.0 M; substrate, 0.25 M.
[b] Moles per mole of compound.
[c] Hydride utilization reached no plateau even after 75 hours.

hydride. From one to three representatives of each class of compound
were used. In all cases where reduction occurred and the product is
indicated, the reaction product after hydrolysis of the boron intermediate
was identified either by gas chromatography or isolation.

The results reveal that aldehydes and ketones are readily reduced to
the alcohol stage. Epoxides are reduced only very slowly — in 38 hours
only 10 to 20 per cent of the estimated hydride utilization is observed.
Carboxylic acids react to form the disiamylboron carboxylate, but no
reduction occurs. Hydrolysis readily yields the original acid.

$$RCO_2H + Sia_2BH \longrightarrow RCO_2BSia_2 + H_2$$
$$RCO_2BSia_2 + H_2O \longrightarrow RCO_2H + Sia_2BOH$$

Presumably the large steric requirements of the disiamylboro group prevents the attack by a second group.

The carbonyl group is still quite reactive to reagents of lower steric requirements, such as diborane and sodium borohydride.

Acid chlorides, acid anhydrides, and esters fail to react. On the other
hand, lactones react readily with the uptake of one hydride per mole of
lactone. Hydrolysis yields the corresponding hydroxyaldehyde in very
good yield. Consequently, the reduction of lactones by disiamylborane
provides a convenient new route to hydroxyaldehydes.

Isopropenyl acetate reacts with the utilization of 2 moles of reagent. However, the product of the reaction has not been identified.

Primary acid amides react with the liberation of 2 moles of hydrogen, but no reduction occurs. Presumably here as in the carboxylic acid case, the large steric requirements of the disiamyl group hinder the reduction. Indeed, it is difficult to visualize the presence of two such bulky groups on the single nitrogen atom of the amide. It would appear more probable that the structure is a derivative of the tautomeric form of the amide.

In the disubstituted amide, reaction with the carbonyl group readily occurs. This is somewhat unexpected, in view of the inertness of esters. Perhaps the ability of the nitrogen atom to place a negative charge on the oxygen of the carbonyl group provides a more reactive, electron-rich center for attack by the electrophilic reagent.

In any event, addition occurs cleanly with the utilization of 1 mole of reagent. Consequently, this provides a new synthetic route for proceeding from the carboxylic acid to the corresponding aldehyde.

Yields of aldehyde in the neighborhood of 80 per cent were demonstrated.

Both the nitrile and the nitro group underwent a very slow reaction, which had failed to reach a plateau after 75 hours. On the other hand, the labile azo grouping is stable to the reagent. Moreover, cyclohexanone oxime reacted to liberate hydrogen but failed to undergo reduction. This has important synthetic implications. It should be possible to protect ketones against the reagent by forming the oxime while operating on another reactive group in the molecule.

The inertness of the oxime is presumably due also to the large steric requirements of the initial reaction product.

Finally, the thiol group, as in thiophenol, fails to liberate hydrogen, the sulfone grouping is inert, and the sulfonic acid group reacts to liberate hydrogen but does not undergo reduction.

Rates of Reduction

While the results of the experiment summarized in Table 18-1 provide a rough measure of the relative rates of reaction of various functional groups with the reagent, it appeared desirable to place this on a more quantitative basis. Accordingly a number of reactions were run at 0° in which the rate of appearance of product or rate of disappearance of disiamylborane was followed. The tetrahydrofuran solution was 0.5 M in each component initially.

In Figure 18-1 is shown the rate of reduction of n-heptaldehyde and benzaldehyde. It is evident that the reaction is very fast, comparable to the rate of reaction of typical terminal olefins, such as 1-hexene, and acetylenes, such as 1-hexyne.

The reduction of ketones is generally much slower and varies considerably with the structure (Figure 18-2). Cyclohexanone is very much faster than the other ketones and approaches the aldehydes in its reactivity. On the other hand, 2-heptanone and acetophenone are much slower. There should be little difficulty in reducing aldehydes selectively in the

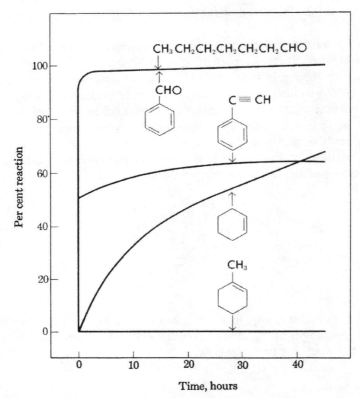

Figure 18-1 Reaction of disiamylborane with representative aldehydes, olefins, and acetylenes in tetrahydrofuran at 0°.

presence of the great majority of ketone structures.   Finally, benzophenone is quite slow.

It would be anticipated that the rate of reduction of a ketone by disiamylborane would be very sensitive to the steric requirements of the ketone. This is shown by the results of a competitive reduction.   With disiamylborane, the relative rate constants $k_{2\text{-heptanone}}/k_{\text{pinacolone}}$ is 12.9, whereas diborane yields the value 1.7.[8]

In Figure 18-3 are shown the results of the rate studies for various carboxylic acid derivatives — ethyl caproate, N,N-dimethylbenzamide and N,N-dimethylcaproamide, and γ-butyrolactone and γ-valerolactone. The reaction with the simple ester is negligible over 40 hours.   As dis-

[8] H. C. Brown and W. Korytnyk, unpublished observations.

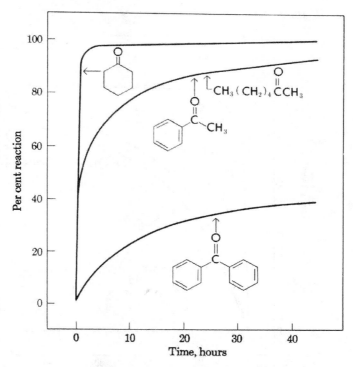

Figure 18-2 Reaction of disiamylborane with representative ketones in tetrahydrofuran at 0°.

cussed earlier, the dimethylamides are far more reactive. Finally, the reactivities of γ-lactones approaches that of the more reactive ketones.

Comparison of Disiamylborane and Diborane

The results reveal both interesting similarities and differences between disiamylborane and diborane.

With both reagents alcohols and phenols evolve 1 mole of hydrogen, but no reaction occurs. Aldehydes and ketone are readily reduced by both reagents, but disiamylborane appears to be much more sensitive to the structure of the ketone. Diborane converts epoxides to the corresponding alcohol at a moderate rate, whereas the reaction with disiamylborane is very slow and may be considered negligible for most practical purposes.

The rate of reduction of carboxylic acids by diborane is amazingly fast, as fast or faster than the reduction of aldehydes. Disiamylborane reacts with the acid to form the disiamylboron acylate, but no reduction occurs.

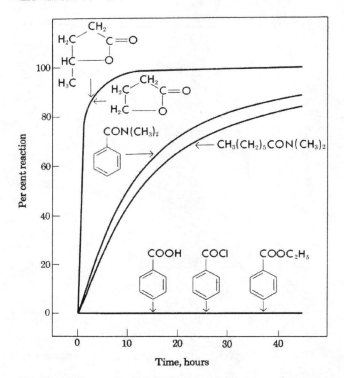

Figure 18-3 Reaction of disiamylborane with representative derivatives of carboxylic acids in tetrahydrofuran at 0°.

Neither reagent reacts at any significant rate with acid chlorides. Esters are reduced at a slow rate by diborane; the corresponding reaction with disiamylborane is negligible.

Both reagents react with primary amides to liberate 2 moles of hydrogen, but no reduction occurs. Diborane reacts with N,N-dimethylamides to utilize 2 moles of hydride. Presumably a reduction to the tertiary amine occurs. On the other hand, only 1 mole of disiamylborane is taken up by such tertiary amides; hydrolysis yields the aldehyde. Diborane reacts readily with oximes to form the corresponding hydroxylamine,[9] whereas disiamylborane reacts only to liberate hydrogen, without reduction occurring. Nitrile groups are rapidly reduced by diborane, whereas the corresponding reaction with disiamylborane is very slow, being incomplete

[9] H. Feuer and B. F. Vincent, Purdue University, private communication.

even after 75 hours at 0°.   The azo linkage is reduced by diborane; disiamylborane fails to react.   No significant reduction of nitro groups by diborane has been observed, whereas a slow reaction occurs with disiamylborane.

Neither reagent reacts with sulfones, but the data are incomplete for diborane in the case of the other sulfur derivatives.   However, it should be very useful that the disulfide linkage is not attacked by disiamylborane.

It should be recalled that the reactivity toward diborane of several representative groups was observed to be[4]: carboxylic acids > olefins > ketones > nitriles > epoxides > esters > acid chlorides.   Disiamylborane is much more sensitive to the structure of the organic component than diborane, whether the reaction under consideration is hydroboration of an olefin (Chapter 13) or reduction of a functional group (Figure 18-2).   Consequently, it is more difficult to define the relative reactivities of representative groups.   With this caution in mind, the data indicate the following approximate order of reactivities: aldehydes > ketones > nitriles > epoxides ≫ esters, acid chlorides, carboxylic acids.

The reactivity of olefins is not indicated in this series, since different structures exhibit widely different reactivities.   However, terminal olefins would fall very near or at the top of the list.   The marked shift in the relative position of the carboxylic acid grouping is noteworthy.   Also important is the observation that so many groupings have relatively low or negligible reactivity toward the reagent, whereas olefin linkages, which are not highly hindered, and acetylene linkages retain very high activity toward disiamylborane.

Selective Reductions and Hydroborations

It is evident that the individual characteristics of disiamylborane provide an additional potent tool for the armory of the synthetic chemist.   This reagent is a mild reducing agent, sensitive to the structure of molecule undergoing reaction, with characteristics quite different from those of either diborane or the complex hydrides.

For example, the reagent would appear to permit the selective reduction of aldehyde or ketone groups in the presence of disulfide, azo, or epoxide structures.   The half-reduction of a lactone to the hydroxyaldehyde in the presence of many ketone groups appears practical.   A dimethylamide group of a dicarboxylic acid could be reduced to the aldehyde in the presence of the free carboxylic acid group.   The ready protection of a ketone grouping through formation of the oxime should also be valuable.

Finally, it is evident that the reagent should be especially valuable for the selective hydroboration of many types of olefinic and acetylenic bonds in the presence of epoxide, carboxylic acid, carboxylic ester, nitrile, disulfide, etc., groupings. For example, it proved possible, by application of the reagent, to convert 10-undecenoic acid into 11-hydroxyundecanoic acid in very high yield.[7]

This question of selective hydroboration of functional derivatives will be considered more fully in Chapter 19.

19 | Hydroboration of Functional Derivatives

As long as the only significant synthetic route to the organoboranes involved the use of reactive organometallics, such as the Grignard reagent, it was understandable that organoboranes with functional groups were very rare. In the case of the benzeneboronic acids, it was possible to utilize relatively standard synthetic methods to introduce such substituents into the aromatic nucleus.[1] However, a search of the literature fails to uncover any functional derivatives of aliphatic organoboranes prior to the application of hydroboration to their synthesis.

The organoboranes are reactive intermediates with steadily growing possibilities for organic syntheses (Chapter 3). The ready synthesis of organoboranes with functional derivatives will, for the first time, make it practical to utilize building blocks containing already incorporated functional groups in synthetic procedures based upon the use of organometallics. It is therefore important to review the scope and possible limitations in the application of the hydroboration reaction to unsaturated derivatives containing representative functional groups.

Up to the present time, relatively little published work in this area has appeared. The available information will be reviewed in three sections: (1) unsaturated derivatives with relatively inert substituents, (2) vinylic and allylic halides, and (3) unsaturated derivatives with reducible groups. Finally, an attempt will be made to indicate the probable scope of the hydroboration reaction applied to unsaturated derivatives containing functional groups.

[1] H. G. Kuivila and A. R. Hendrickson, *J. Am. Chem. Soc.*, 74, 5068 (1952).

Derivatives with Inert Substituents

The hydroboration of olefins containing relatively inert substituents appears to proceed without difficulty. Thus *p*-chlorostyrene and *p*-methoxystyrene are readily converted into the corresponding organoboranes and the latter oxidized to the alcohols.[2] Except for a significant change in the position taken by the entering boron atom, the reactions appear to differ insignificantly from the corresponding hydroboration-oxidation of styrene.

Similarly, the hydroboration of anethole and *cis-trans*-2-*p*-anisyl-2-butene proceeds without complications.[3] The same influence of the *p*-methoxy group in altering the distribution from that in *trans*-1-phenyl propene[2] is evident.

This influence of the substituent on the direction of hydroboration is evident in the aliphatic series. Thus the hydroboration-oxidation of

[2] H. C. Brown and G. Zweifel, *J. Am. Chem. Soc.*, **82**, 4708 (1960).
[3] E. L. Allred, J. Sonnenberg, and S. Winstein, *J. Org. Chem.*, **25**, 26 (1960).

vinyltrimethylsilane proceeds without difficulty but yields 63 per cent of the primary, 37 per cent of the secondary alcohol.[4] It may be recalled that the analogous carbon compound, t-butylethylene, yields a 94:6 distribution of the primary and secondary alcohols.[2]

Similarly, allyltrimethylsilane and allyltrichlorosilane have been converted to the corresponding organoboranes.[5]

$$3(CH_3)_3SiCH_2CH{=}CH_2 \xrightarrow{\text{HB}} [(CH_3)_3SiCH_2CH_2CH_2]_3B$$

$$3Cl_3SiCH_2CH{=}CH_2 \xrightarrow{\text{HB}} (Cl_3SiCH_2CH_2CH_2)_3B$$

Again, there is no mention of any unusual directive effects, although such might be anticipated for compounds containing such electronegative substituents.

Allylmethyl thioether was converted to the organoborane and protonolyzed with propionic acid to form methyl-n-propylsulfide.[6] This reaction was of interest in demonstrating the feasibility of the hydroboration-protonolysis procedure as a means of achieving saturation of carbon-carbon double bonds in molecules containing relatively reactive sulfur substituents.

$$3CH_3SCH_2CH{=}CH_2 \xrightarrow{\text{HB}} (CH_3SCH_2CH_2CH_2)_3B$$

$$\downarrow RCO_2H$$

$$3CH_3SCH_2CH_2CH_3$$

[4] D. Seyferth, *J. Inorg. & Nuclear Chem.*, **7**, 152 (1958).

[5] B. M. Mikhailov and T. A. Shchegoleva, *Izvest. Akad. Nauk. SSSR*, **1959**, 546.

[6] H. C. Brown and K. Murray, *J. Am. Chem. Soc.*, **81**, 4108 (1959).

Finally, the organoborane from 4-chloro-1-butene has been prepared[7] and distilled[8] (b.p. 126 to 128° at 10^{-5} mm) without evidence of any chemical change.

Hawthorne has recently achieved the synthesis of two interesting heterocycles via the reaction of trimethylamine-t-butylborane with divinyl ether and dimethyldivinylsilane.[9]

$$
O\begin{matrix} CH{=}CH_2 \\ \\ CH{=}CH_2 \end{matrix} + (CH_3)_3N{:}BH_2C(CH_3)_3 \longrightarrow O\begin{matrix} CH_2CH_2 \\ \\ CH_2CH_2 \end{matrix}BC(CH_3)_3 + N(CH_3)_3
$$

$$
(CH_3)_2Si\begin{matrix} CH{=}CH_2 \\ \\ CH{=}CH_2 \end{matrix} + (CH_3)_3N{:}BH_2C(CH_3)_3 \longrightarrow (CH_3)_2Si\begin{matrix} CH_2CH_2 \\ \\ CH_2CH_2 \end{matrix}BC(CH_3)_3 + N(CH_3)_3
$$

The hydroborations of vinylic and allylic halides possess unique characteristics, which makes it desirable to consider them as a special class.

Derivatives with Vinylic and Allylic Halogen

The hydroboration of vinyl chloride proceeded to yield thermally unstable products which could not be separated. By treating vinyl chloride at −80° in the presence of dimethyl ether, the authors observed vigorous exothermic reactions on warming up. They isolated small quantities of a material which they identified as an etherate of β-chloroethylboron dichloride.[10] Treatment of this material with water produced boric acid, hydrogen chloride, and ethylene.

$$
R_2O{:}BCl_2CH_2CH_2Cl + 3H_2O \longrightarrow CH_2{=}CH_2 + B(OH)_3 + 3HCl + R_2O
$$

[7] M. F. Hawthorne, *J. Am. Chem. Soc.*, **82**, 1886 (1960).
[8] P. Binger and R. Köster, *Tetrahedron Letters*, No. 4, 156 (1961).
[9] M. F. Hawthorne, *J. Am. Chem. Soc.*, **83**, 2541 (1961).
[10] M. F. Hawthorne and J. A. Dupont, *J. Am. Chem. Soc.*, **80**, 5830 (1958).

This reaction is quite similar to the behavior of β-chloroethyl derivatives in the silicon compounds.[11]

$$Cl_3SiCH_2CH_2Cl + 4H_2O \longrightarrow CH_2{=}CH_2 + Si(OH)_4 + 4HCl$$

The authors suggest that the violent exothermic reactions observed in the hydroboration of vinyl chloride may well result from related changes.

$$B(CH_2CH_2Cl)_3 \longrightarrow ClB(CH_2CH_2Cl)_2 \longrightarrow Cl_2BCH_2CH_2Cl$$
$$+ CH_2{=}CH_2 \qquad\qquad + CH_2{=}CH_2$$

p-Chlorostyrene may be considered a phenyl homolog of vinyl chloride. The p-chloro substituent in styrene strongly influences the boron atom to add at the nearer secondary position of the side chain. It would appear dangerous, therefore, to assume that the hydroboration of vinyl chloride must proceed only to give the β-chloroethyl derivative. It is much more probable that the reaction involves attachment of boron to both carbon atoms of the double bond, and essentially nothing is known of the chemical characteristics of derivatives having both boron and chlorine attached to the same carbon atom. This would appear to be an ideal system for carbene formation.

Evidence for such a directive effect is evident in the hydroboration of 3-chlorocyclohexene.[8] The product is reported to be exclusively the 2-haloalkylborane, which undergoes rapid elimination of the kind postulated above.

[11] L. H. Sommer, G. M. Goldberg, E. Dorfman, and F. C. Whitmore, *J. Am. Chem. Soc.*, **68**, 1083 (1946).

It may be recalled that the hydroboration of 3-methylcyclohexene proceeds to place the boron almost equally on the two carbon atoms of the double bond[12] (Chapter 7).

Attention might be called here to the violent reactions realized in the high-temperature reactions of diborane and tetrafluoroethylene.[13] Even at lower temperatures, the related reactions of diborane with tetrafluoroethylene, trifluoroethylene, 1,1-difluoroethylene, and vinyl fluoride proved to be difficult to control and led to complex mixtures.[14]

The hydroboration of allyl chloride proceeded quite smoothly.[10] Distillation yielded tri-(γ-chloropropyl)-borane (b.p. 118° at 1 mm) and di-(γ-chloropropyl)-boron chloride (b.p. 76° at 1 mm). Upon treatment with aqueous base, these compounds underwent cyclization to give nearly quantitative yields of cyclopropane. In this case also the reaction resembles closely the analogous formation of cyclopropane from γ-chloropropylsilicon trichloride.[15]

The hydroboration of substituted allyl chlorides, followed by treatment with alkali, has been suggested as a general synthesis of substituted cyclopropanes.[7]

[12] H. C. Brown and G. Zweifel, *J. Am. Chem. Soc.*, **83**, 2544 (1961).

[13] F. G. A. Stone and W. A. G. Graham, *Chem. & Ind. (London)*, **1955**, 1181.

[14] B. Bartocha, W. A. G. Graham, and F. G. A. Stone, *J. Inorg. & Nuclear Chem.*, **6**, 119 (1958).

[15] L. H. Sommer, R. E. Van Strien, and F. C. Whitmore, *J. Am. Chem. Soc.*, **71**, 3056 (1949).

It has been suggested that the modest yield of tri-(γ-chloropropyl)-borane realized by Hawthorne and Dupont is the result of the influence of the chlorine substituent in directing as much as 30 per cent of the boron to the secondary position.[8]   The resulting structure, 1-chloro-2-propyl boron derivative, would eliminate rapidly, forming propylene (which would undergo further hydroboration) and a boron chloride linkage.

The use of a dialkylborane greatly reduces this side reaction.[8,16]

Derivatives with Reducible Groups

Acrylonitrile and methyl methacylate were early treated with diborane in the absence of ether solvents.   A slow reaction was observed, but the products were complex and not completely separated nor characterized.[17]

[16] H. C. Brown and K. Keblys, unpublished observations.

[17] F. G. A. Stone and H. J. Eméleus, *J. Chem. Soc.* (*London*), **1950**, 2755.

The ester grouping is reduced relatively slowly by diborane (Chapter 17). Therefore it should offer no major difficulty in the hydroboration of unsaturated esters. For example, Fore and Bickford carried out the hydroboration of methyl oleate to form the corresponding tris-(carbo-methoxyalkyl)-borane. Oxidation by alkaline hydrogen peroxide provided 9- and 10-hydroxyoctadecanoic acids in equimolar amounts.[18]

Similarly, Dulou and Chrétien-Bessière converted methyl undecenoate to methyl 11-hydroxyundecanoate.[19]

The extension of the hydroboration to the sugar area by Wolfrom and Whitely is worthy of note.[20] In this case the reducible acetal linkage was protected as the isopropylidine derivative, and the hydroboration-oxida-

[18] S. P. Fore and W. G. Bickford, *J. Org. Chem.*, **24**, 920 (1959).

[19] R. Dulou and Y. Chrétien-Bessière, *Bull. soc. chim. France*, **1959**, 1362.

[20] M. F. Wolfrom and T. E. Whiteley, Abstracts of Papers, 137th Am. Chem. Soc. Meeting, Lec. 2–D.

tion procedure transformed 1,2-*O*-isopropylidene-4-vinyl-α-*D*-xylo-tetro-furanone into 5-deoxy-1,2-*O*-isopropylidene-α-*D*-xylo-hexofuranose.

The carboxylic acid group reacts exceedingly rapidly with diborane. Consequently, the hydroboration of undecenoic acid to form the corresponding organoborane proceeds with concurrent reduction of the acid group. However, as pointed out in Chapter 18, disiamylborane does not reduce carboxylic acid groups. Its use permits the simple preparation of 11-hydroxyundecanoic acid in a yield of 82 per cent.[21]

$$
\begin{array}{ccc}
CH_2 & & CH_2\!-\!BSia_2 \\
\| & & | \\
CH & & CH_2 \\
| & +\ 2Sia_2BH\ \longrightarrow & | & +\ H_2 \\
(CH_2)_8 & & (CH_2)_8 \\
| & & | \\
CO_2H & & CO_2BSia_2
\end{array}
$$

$$\Big\downarrow [O]$$

$$
\begin{array}{c}
CH_2OH \\
| \\
(CH_2)_9 \\
| \\
CO_2H
\end{array}
$$

One problem should be mentioned. The direct hydroboration of oleic and undecenoic esters offers no difficulty — the ester groupings are far removed from the reaction center. However, an attempt to apply the same reaction to ethyl vinyl acetate results in a reduction of the ester grouping that is competitive with the hydroboration of the double bond.[16] Presumably this difficulty arises because one of the intermediates in the hydroboration reaction is in position to react with the ester group via a cyclization mechanism.

[21] H. C. Brown and D. B. Bigley, *J. Am. Chem. Soc.*, **83**, 486 (1961).

It is possible to avoid this difficulty by use of the monofunctional hydro-borating agent, disiamylborane.[16]

$$CH_2{=}CHCH_2CO_2Et \xrightarrow{\text{Sia}_2BH} Sia_2BCH_2CH_2CH_2CO_2Et$$

$$\downarrow [O]$$

$$HOCH_2CH_2CH_2CO_2Et$$

Thus far no report has been published on the hydroboration of other unsaturated compounds containing functional groups. Steroids have been hydroborated in which the carbonyl functions have been protected as the cyclic acetals.[22]

It also appears that oximes should provide a convenient means of pro-tecting carbonyl groups in hydroborations with disiamylborane.

Scope

The research thus far published does not permit an exact definition of the full scope of the hydroboration of functional derivatives. However, the trialkylborane grouping would appear capable of tolerating many types of substituents. Halogens in the beta position to boron appear to be eliminated rapidly. Strong oxidizing groups, such as peracids, hypo-chlorites, and peroxides, are obviously out of the question. However, most other structures, such as aldehyde and ketone, epoxide, carboxylic acid, ester and amide, nitrile and nitro, would appear to be compatible with the boron-carbon linkage in the organoborane. They should offer no

[22] M. Nussim and F. Sondheimer, *Chem. & Ind.* (*London*), 1960, 400.

major difficulties at the mild temperatures generally utilized for the hydroboration reaction and most of the subsequent conversion reactions of the organoboranes (Chapter 3).

In the hydroboration stage only the aldehyde and some of the ketone groups would appear to offer major difficulties. However, these groups are readily protected as the acetals, and possibly as the oximes. We may therefore look forward to a rapid extension of the hydroboration reaction to the functional derivatives.

20 | Epilog

All pleasant tasks must come to an end. It has been a real pleasure to have had this opportunity to review developments within the last few years. Indeed, it is difficult to realize that this entire development had its inception only a few short years ago.[1] It is even more startling that this development was the result of a minor experimental discrepancy which might have been shrugged off into oblivion with an easy conscience.

At the time Dr. B. C. Subba Rao was exploring the enhanced reducing action of sodium borohydride under the influence of aluminum chloride.[2] He was following the reaction by adding a standard solution of the reagent to weighed samples of representative organic compounds. After standard intervals of time, aqueous hydrochloric acid was added to transform the residual "hydride" into hydrogen, which was collected and measured. The decrease in the amount of hydrogen obtained from that realized in blank determinations gave us the moles of hydride utilized per mole of compound.

In this manner he established that representative aldehydes and ketones utilized one "hydride" per mole. Obviously reduction was occurring to the alcohol stage. Nitriles utilized 2 moles of "hydride," corresponding to reduction to the amine. Esters, such as ethyl acetate and ethyl stearate, revealed the uptake of 2 moles of hydride. Evidently, reduction to the alcohol was occurring here also. One of the compounds Dr. Subba Rao examined was ethyl oleate. It showed a "hydride" uptake of 2.4.

Dr. Subba Rao is a most productive laboratory worker. He came to me with the data for many, many compounds. The discrepancy was noted and discussed. Dr. Subba Rao suggested that the ethyl oleate might not

[1] H. C. Brown and B. C. Subba Rao, *J. Am. Chem. Soc.*, **78**, 5694 (1956).
[2] H. C. Brown and B. C. Subba Rao, *J. Am. Chem. Soc.*, **78**, 2582 (1956).

have been pure — the difficulties of getting pure samples of oleic acid are well known.

We discussed the possibility of dropping the experiment. After all, this was merely one odd result in several hundred experiments. Fortunately, the research director is in an enviable position to insist on high standards — he does not have to do the experimental work. It was recommended that Dr. Subba Rao repeat the experiment with a purified sample.

This yielded the same result! The original sample of ethyl oleate had been pure. It required only a little more effort to establish that we were achieving the simultaneous reduction and hydroboration of ethyl oleate.

How is it possible that a reaction of such generality was not uncovered and utilized long ago?

As I look back from my present knowledge, it appears to me that we should have recognized the reaction much earlier. In 1936 I was beginning my doctorate research in the area of diborane chemistry. I was cautioned not to use stopcocks in any area of the vacuum line exposed to diborane vapor. Stopcock grease became very gummy when exposed to the vapor. At that time the stopcock greases had a rubber base. Later, when the Apiezon greases, prepared from a petroleum base, became available, the difficulty disappeared. I recall thinking about the phenomenon, but did not put two and two together and recognize that diborane must be cross-linking the unsaturated centers in the rubber-base grease.

Again, some time later, we utilized trimethylborane as a means of obtaining thermodynamic data for the dissociation of addition compounds.[3] We encountered no difficulty in purifying and storing trimethylborane. However, later, when I attempted to utilize triethylborane for similar studies, I encountered major difficulties. A sample of triethylborane would be purified thoroughly. It would be placed in storage while other reactants were purified. But when the time came to use it, the material was impure. It appeared to evolve small quantities of ethylene. Today, I should interpret the reaction as involving a reversible dissociation of the triethylborane into diethylborane and ethylene.

$$(C_2H_5)_3B \xrightleftharpoons{\Delta} (C_2H_5)_2BH + C_2H_4$$

Finally, triethylborane was used only in qualitative estimates of the base strengths of amines.[4]

[3] H. C. Brown, M. D. Taylor, and M. Gerstein, *J. Am. Chem. Soc.*, **66**, 431 (1944).

[4] H. C. Brown, *J. Am. Chem. Soc.*, **67**, 374 (1945).

When one is young, he is impatient with these little discrepancies and tends to hurry past them to the main objective. As one gains experience, he appreciates better the importance of these discrepancies, but professional obligations render it more and more difficult to maintain direct contact with experimental observations, where there is an opportunity to recognize these nuggets.

It is well recognized that a research program such as is summarized in this book depends upon the ability and enthusiasm of many co-workers. I am indeed indebted to a most capable group of graduate students and post-doctorates, whose individual contributions are indicated in the text. However, it will be recognized by any reader that two of my associates have made truly staggering contributions to this research program — Dr. B. C. Subba Rao and Dr. George Zweifel. I am confident that the reader will wish to become personally acquainted with these unusually productive co-workers. Accordingly, their pictures and biographies are presented in the pages immediately following. The hydroboration program is indeed indebted to their magnificent contributions.

At this point it is tempting to take up one's crystal ball and predict future developments. Unfortunately, the academic profession has been educated to the value of new chemistry, and there are obvious dangers in making such predictions. Accordingly, I shall content myself with the observation that progress has been very rapid in the past few years and shows no signs of deceleration.

Finally, in conclusion, may I present the moral of this volume —

"Tall oaks from little acorns grow"

About B. C. Subba Rao

Bookinkere Channakeshaviah Subba Rao was born in Mysore, India, on December 8, 1923, the third child of Professor B. Channakeshaviah. He attended Mysore University in Bangalore, where he received the B.Sc.(Hons.) in chemistry from Central College in 1944, standing first in chemistry for the University. In 1946 he received the M.S. degree (with Distinction) from the same university. He then joined the Indian Institute of Science at Bangalore, transferring in 1951 to the teaching staff of the new Indian Institute of Technology, Kharagpur (near Calcutta).

The following year he entered Purdue, receiving his Ph.D. degree in 1955. The following two years were devoted to postdoctorate research with Professor H. C. Brown in the areas of selective reductions and hydroborations. In 1957 he returned to India, becoming Senior Scientific Officer in the Organic Chemistry Division of the National Chemical Laboratory of India and continues in that position. At the Laboratory he is engaged mainly with research problems having an applied bias, such as the utilization of indigenous raw materials and the development of new processes for the production of industrial chemicals. He is a recognized Ph.D. guide of Poona University and directs the research of several Ph.D. candidates.

In 1946 he married Uma Chandrasekhar. They have three children, a boy Ashok, born in Bangalore in 1948, a girl Sudha Rao, born in West Lafayette, Indiana, in 1957, and a boy Somesh, born in Poona in 1958.

Dr. Subba Rao is a member of the American Chemical Society, a member of Sigma Xi, an associate member of the Royal Institute of Chemistry (London), and a Fellow of the Indian Chemical Society.

About George Zweifel

George Zweifel was born on October 2, 1926, in Rapperswil, Switzerland, the second child of a family of seven. He attended Minerva College in Zurich, entering the Swiss Federal Institute (E.T.H.) in 1948 to study agricultural chemistry. He obtained the diploma of engineer in 1952, and then began work on his doctorate thesis with Professor H. Deuel, being appointed to the rank of instructor in 1953. He was granted the degree Dr. Sc. Techn. in 1955.

The award of an exchange fellowship made it possible for him to spend the following year with Professor E. L. Hirst at the University of Edinburgh, where his main research interest dealt with conformational problems in the carbohydrate field. The period 1956–1958 was likewise devoted to problems in sugar chemistry, working with Professor M. Stacey at the University of Birmingham. He came to Purdue in 1958, undertaking research in the new area of hydroboration chemistry. He became Professor Brown's personal research assistant in 1960, with the title of Research Associate.

George Zweifel married Johanna Staub in 1953. He is the father of two boys, Stephen and Hanspeter, both born in West Lafayette.

He is a member of the American Chemical Society and Sigma Xi.

B. C. SUBBA RAO

GEORGE ZWEIFEL

H. C. BROWN

About
the Author

Herbert C. Brown was born in London, England on May 22, 1912. He came to the United States at the age of 2 and all his early education was obtained in Chicago schools. He attended Wright Junior College, 1934–1935 (Assoc. Sci., 1935) and the University of Chicago (B.S., 1936; Ph.D., 1938). His doctorate thesis was carried out under the direction of Professor H. I. Schlesinger in a promising area, "The Reaction of Diborane with Organic Compounds Containing a Carbonyl Group." The following year was spent with Professor M. S. Kharasch and resulted in the discovery of the peroxide-catalyzed chlorination of aliphatic derivatives by sulfuryl chloride, the chlorocarboxylation reaction, and studies in the mechanisms of free radical reactions.

In 1939 he joined the staff as personal assistant to Professor H. I. Schlesinger. Academic work soon gave way to emergency research on new volatile compounds of uranium and the field generation of hydrogen. This work led to practical new synthetic methods for the synthesis of diborane and metal borohydrides. At the same time the chemistry of boron was devoted to the initiation of a new research approach to the role of steric effects in chemistry.

In 1943 he transferred to Wayne University as assistant professor, becoming associate professor in 1946. He was invited to Purdue University in 1947 with the rank of Professor, and became the R. B. Wetherill Professor in 1959, and the R. B. Wetherill Research Professor in 1960. At Purdue he has pursued research programs in both organic and inorganic chemistry, involving the study of molecular addition compounds, the chemistry of steric strains, directive effects in aromatic substitution, selective reductions, and hydroboration.

He married a fellow chemistry student at the University of Chicago, Sarah Baylen, in 1937. They have one son, Charles Allan, born in Detroit in 1943 and now studying chemistry at Purdue.

Professor Brown's contributions have been recognized in numerous awards and honorary lectures, among which may be mentioned: Harrison Howe Lecturer, 1953; Centenary Lecturer, 1955; National Academy of Sciences, 1957; Nichols Medallist for 1959; A.C.S. Award for Creative Work in Synthetic Organic Chemistry, 1960; Honorary Member of Phi Lambda Upsilon, 1961.

Index

The author wishes to acknowledge the invaluable assistance of Drs. George Zweifel and Brian J. Garner in proofreading the text and in the preparation of the index.